The 1924 Coolidge-Dawes Lincoln Tour

A cross-country road rally of 100,000 automobiles.
A continuous political parade from the Atlantic to the Pacific.
A story about the famous Lincoln Highway in its early years.

Larry L. Krug

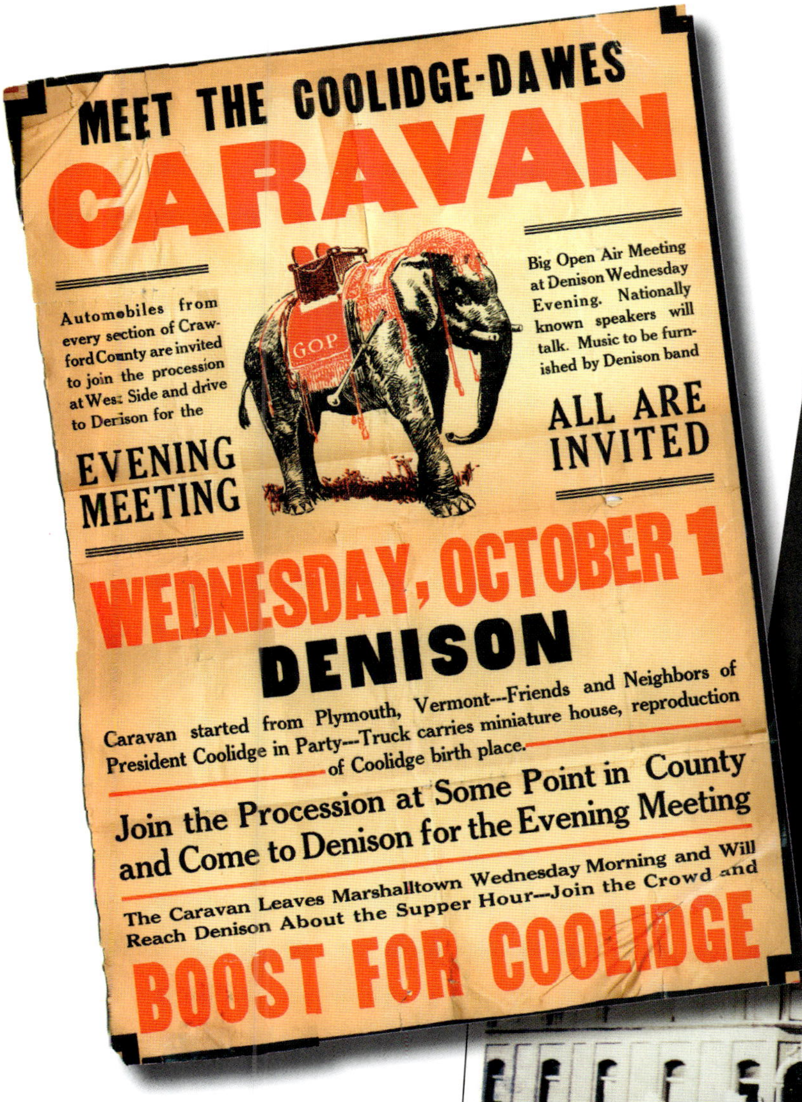

Published by Schiffer Publishing Ltd.
4880 Lower Valley Road
Atglen, PA 19310
Phone: (610) 593-1777; Fax: (610) 593-2002
E-mail: Info@schifferbooks.com

For the largest selection of fine reference books on this and related subjects, please visit our web site at **www.schifferbooks.com**
We are always looking for people to write books on new and related subjects. If you have an idea for a book please contact us at the above address.

This book may be purchased from the publisher.
Include $3.95 for shipping.
Please try your bookstore first.
You may write for a free catalog.

In Europe, Schiffer books are distributed by
Bushwood Books
6 Marksbury Ave.
Kew Gardens
Surrey TW9 4JF England
Phone: 44 (0) 20 8392-8585; Fax: 44 (0) 20 8392-9876
E-mail: info@bushwoodbooks.co.uk
Website: www.bushwoodbooks.co.uk
Free postage in the U.K., Europe; air mail at cost.

Library of Congress Control Number: 2007926610

Type set in New Baskerville

ISBN: 978-0-7643-2763-6
Printed in China

Introduction

Welcome to the Coolidge-Dawes caravan, commonly called "The Coolidge-Dawes Lincoln Tour," as it followed the route of the Lincoln National Highway (now basically U.S. 30) for a large portion of its journey. Perhaps the most aggressive campaign project ever conceived for that era, the caravan officially rolled out of Plymouth, Vermont, the birthplace of Calvin Coolidge, at 12 noon on September 9, 1924 on a 56-day speaking rally through New Hampshire, Massachusetts, New York, New Jersey, Pennsylvania, and westward to the Pacific coast, where it would visit California, Oregon, and Washington. Although large areas of the Lincoln Highway were still unpaved – ranging from dirt in good weather to mud in bad weather – the 1924 caravan moved on, eventually involving well over 100,000 vehicles...in some ways truly the first major cross-country road rally!

Hundreds of speaking programs, often accompanied with pageantry and excitement, were presented along the tour path. They involved local and state dignitaries, plus a small core of speakers who made the entire trip. The plan was simple. State organizations would meet the caravan at their respective state border and escort the parade across their state, stopping at cities and towns, large and small. At the state boundary they would turn the caravan over to officials and their accompanying caravan vehicles from the next state. In high density areas the parade often reached several miles long. Going through the plains of Nebraska and the snow-covered high areas of Wyoming, it dwindled to only a few brave automobiles. A Reo truck decorated with photos of Coolidge and Dawes and of Plymouth Notch led the tour the entire distance, and also functioned as the stage as it carried the amplifying equipment. As the caravan approached towns, be it morning, afternoon, or night, there was a festive atmosphere. Horns honked; there were police escorts and school marching bands, local political groups, pennants, and flags. Schools were released so school children could line the curbs. Houses along the route often flew the American flag and were decorated in red, white and blue bunting.

A register was carried throughout the tour to gather signatures of elderly citizens who had voted for Abraham Lincoln. Seldom was a stop made where there wasn't a small group of these men, in their 80s and 90s, who had often traveled miles to come see the caravan and to sign the register. Before the tour was completed, a second volume of the register had to be added. The register was later presented to President Coolidge.

Although neither Calvin Coolidge nor Charles Dawes actually participated in the tour, it undoubtedly was one of the most effective campaign public relations projects of the campaign. As an original departure in electioneering it was one of the most unique features ever attempted in a political campaign. Media covered the hard hitting speeches that were delivered day after day in support of the Republican ticket. Although speeches were delivered by many different politicians, there were reoccurring themes. John W. Davis, the Democratic candidate, was seldom mentioned, although Charles Bryan, his running mate and brother to William Jennings Bryan, was bitterly denounced in speech after speech, sharing the dubious honor with the dire warnings of the Robert LaFollette ticket.

Over five million people turned out to witness the caravan as it rolled along 6,500 miles through 17 states. Over 300,000 attended one or more of the 400 meetings at the various stops. In Westchester County, New York over 2,000 cars joined the parade for that one event alone – decorated in bunting, with flags flying and horns honking. Colorful posters, pennants, and over two million campaign buttons were distributed in route. Chester and Lancaster Counties in Pennsylvania reported the procession to be over 15 miles long with most of the cars decorated and filled with supporters. The local press along the route loved the event, proclaiming it "the largest continuous procession of any kind in all history." Even the tour promoters were caught up in the excitement, stating at its conclusion that "this has been the longest procession in history and is the first instance that a trans-continental trip has been made by motor vehicles on a fixed schedule and arrived on time."

This book is compiled completely from first-hand, eye-witness accounts of the Coolidge-Dawes Lincoln Tour. Mr. E. A. Seidel, who drove the REO Speedwagon lead vehicle the entire length of the trip, kept a daily log. This information, along with the actual accounts of reporters from dozens of newspapers from towns and cities along the caravan route comprise the day by day chronology of the 56-day event. Minor inaccuracies in the news reporting and the misspelling of names have not been corrected but are printed as reported to convey the flavor of the report. The general themes and strategies of the campaign come through clearly as they are reported over and over by the ever "new" local reporters as the campaign moved westward day after day.

Very little is documented about this 1924 political extravaganza and what little is written is often inaccurate. In addition to the newspaper reports we are very fortunate to have so many visuals and mementoes documenting and illustrating the excitement of this trip – for 1924 truly an experiment in modern-day campaigning.

Larry L. Krug

THE HOME TOWN COOLIDGE CLUB

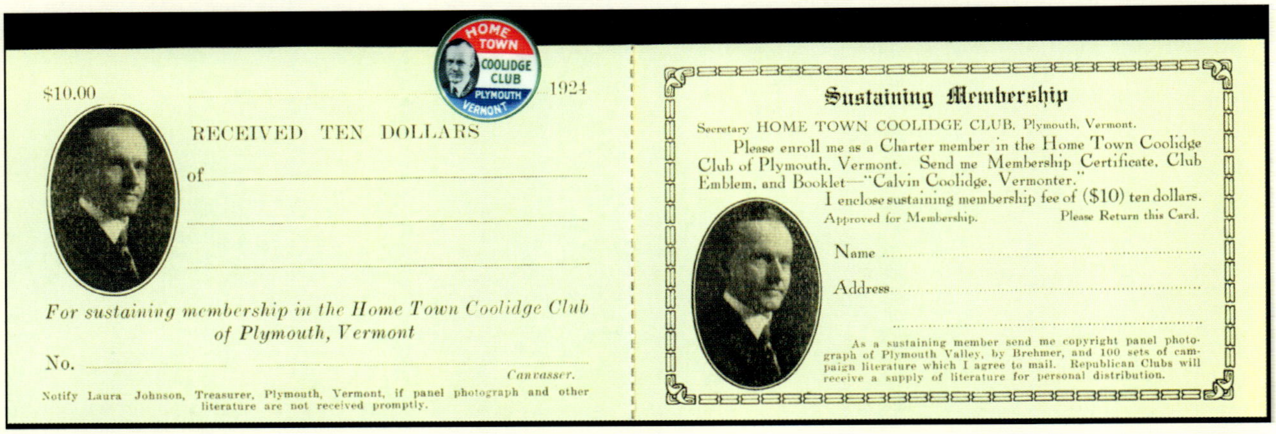

Home Town Coolidge Club membership card and button

Although Calvin Coolidge had been President of the United States for eight months, since the death of President Harding, he still was not well known by the time the 1924 campaign began. Adding to this, those who knew him best felt the media was not depicting an accurate picture of Coolidge.

Those Plymouth, Vermont, friends and neighbors of Coolidge naturally decided to do what Vermonters do – they would set the record straight.

In the early spring of 1924 in the building where President Coolidge was born, a hall just above the room where he first saw the light of day, about 50 residents of Plymouth met and organized the Home Town Coolidge Club. They had officers and set up an Executive Committee. Memberships were promoted at $10 each for a Sustaining Member and $1 for a Charter Member. Materials were produced – membership certificate and emblem, a booklet "Calvin Coolidge, Vermonter," and a Home Town Coolidge Club button. The Club heavily promoted their Official Campaign Song: "Keep Cool and Keep Coolidge." (In 1924, "cool" did not mean hip. It meant the opposite: steady, dependable and calm.) A large photo brochure depicted "The Boyhood Training of President Coolidge" with photos of Coolidge from his birthplace to the Executive Office of the White House; it was a major promotion piece in securing memberships.

The Home Town Coolidge Club also became the "brainchild" behind the idea of the Coolidge-Dawes Lincoln Tour. As described in printed literature in 1924, "The Coolidge-Dawes Lincoln Tour is unique in the annals of national politics. It was conceived by the boyhood friends and neighbors of President Coolidge, all residents of Plymouth, Vermont, and its environs, for the primary purpose of acquainting the voters of the country with the Calvin Coolidge they know. There, 'midst humble setting, began the career of America's Chief Executive, and there, appropriately enough, began the tour that before it is ended will have traversed seventeen states and visited about 300 cities and towns. It derives its name from the fact that almost throughout it will travel the transcontinental highway named for the Great Emancipator."

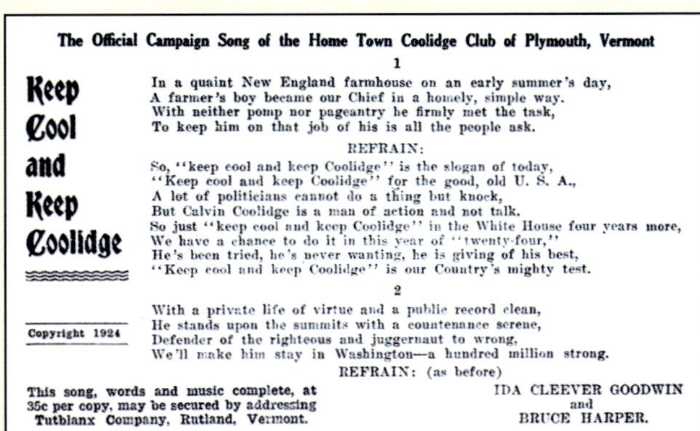

The Official Campaign Song of the Home Town Coolidge Club of Plymouth, Vermont

Keep Cool and Keep Coolidge

1

In a quaint New England farmhouse on an early summer's day,
A farmer's boy became our Chief in a homely, simple way.
With neither pomp nor pageantry he firmly met the task,
To keep him on that job of his is all the people ask.

REFRAIN:

So, "keep cool and keep Coolidge" is the slogan of today,
"Keep cool and keep Coolidge" for the good, old U. S. A.,
A lot of politicians cannot do a thing but knock,
But Calvin Coolidge is a man of action and not talk.
So just "keep cool and keep Coolidge" in the White House four years more,
We have a chance to do it in this year of "twenty-four,"
He's been tried, he's never wanting, he is giving of his best,
"Keep cool and keep Coolidge" is our Country's mighty test.

2

With a private life of virtue and a public record clean,
He stands upon the summits with a countenance serene,
Defender of the righteous and juggernaut to wrong,
We'll make him stay in Washington—a hundred million strong.

REFRAIN: (as before)

Copyright 1924

This song, words and music complete, at 35c per copy, may be secured by addressing Tutblanx Company, Rutland, Vermont.

IDA CLEEVER GOODWIN
and
BRUCE HARPER.

Official Home Town Coolidge Club campaign song

Tuesday, September 9
Day 1

The caravan officially started from Plymouth, Vermont at 12 noon.

Log: Stops at Escausnyville, Claremont, Newport, Keene, New Hampshire and Greenfield, Massachusetts. 134 miles. Spent night at Walden Hotel, Greenfield.

Review: After being presented with the Vermont State flag by the Home Town Coolidge Club, speeches were made by Congressman Jefferies and Senator Fess. Then with a cheer that echoed through the hillside we started on our long journey toward the Pacific Coast, in a driving rainstorm. A better idea of the start can be gathered by reading the Boston Transcript story.

An additional speaker was Capt. John Winant, nominee for Governor of New Hampshire, who spoke at Claremont and Keene.

Hock, driving Ford, burns out a bearing at Woodstock and has a flat tire to change in the pouring rain.

The flag of Massachusetts was presented at Greenfield, in the pouring rain, by the Women's Republican Club.

Average speed: 31.6 miles per hour.

—F. A. Seidel

Coolidge-Dawes Caravan ribbons

Boston Transcipt article:

The Coolidge-Dawes caravan, whose route is to carry it through 300 cities and towns from Plymouth village to the Pacific coast, started its coast to coast trip today in a downpour of rain. A scheduled outdoor meeting was transferred to the village church, across the street from the homestead of John C. Coolidge, the President's father, where about 50 citizens gathered to hear the opening address of Senator Simeon D. Fess of Ohio and former Congressman A. Jefferies of Nebraska, urging the election of the Republican candidates in the coming national election.

Attending the rally in Plymouth today were 19 relatives of the President, forming nearly 30 percent of the gathering in this little village. Five of those present had voted for Abraham Lincoln. One of the latter, Charles Brown, 83 years of age, walked three miles in the rain, over muddy roads, to be present.

The caravan entered New Hampshire from Claremont, Vermont, and Keene was the first city in New Hampshire to welcome it. Although drenched to the skin by heavy rain, hundreds of men and women listened to the speakers who were introduced by Mayor Robert T. Kingsbury. The speakers were Senator Simeon Fess of Ohio, former Congressman A. W. Jeffries of Nebraska, Capt. John C. Winant, Republican candidate for governor of New Hampshire, and Congressman Edward H. Waxon of Nashua, New Hampshire. Other members of the party were State Chairman Dwight S. Hall of Dover and Senator Henry W. Keys, New Hampshire.

When the party crossed the state line from Vermont to New Hampshire Mrs. John G. Glesanes of Bethlehem, national Republican committeewoman, presented former Congressman Jeffries, the chief crusader of the caravan, with the state flag which will be carried on the tour.

The caravan left Keene for Greenfield at 7:30 p.m.

When the Coolidge-Dawes tourists started out from Plymouth to make their long journey across the continent they were accompanied by one of the President's boyhood friends who proposed to make the trip solely as an observer and not to take any active part in the various meetings held along its route.

Herb Moore was the man. But his determination to be seen and not heard was short lived. Early in the trip when somebody in the Caravan's audience asked whether President Coolidge was a real dirt farmer it was more than Moore could stand. He mounted the rostrum and volunteered to answer the question. At that moment an orator was revealed. The audience stood in rapt attention until he had finished and then gave him an ovation. He continued talking at the caravan's meetings from then on throughout the length of the tour.

Moore knew Calvin Coolidge well. As boys they played together and both did all the chores around their respective homes as well as some of the heavier farming. It was of this phase of the President's life that Moore talked and he did it in language easily understood. He was a farmer himself.

So, at practically every stop Herb Moore, boyhood friend of Coolidge, devoted his time to reciting the humbleness and simplicity of the Coolidge home and the town of Plymouth, Vermont. He would talk about Plymouth's population of 29 people, four residences, a general store and a cheese factory. When President Coolidge was appraised of the death of President Harding in his father's home, it had no telephone or electric light. The President's father, Herb would explain, was a stalwart American and served as Justice of the Peace for his little village of Plymouth. The caravan idea, Herb would say, was conceived in Plymouth. It consists of a delivery truck carrying the banner of brass tacks and common sense, a replica of the birth place of the president, and several cars for the speakers.

The Coolidge-Dawes Caravan lead car in front of the Coolidge homestead in Plymouth, Vermont (newsprint)

Colonel John Coolidge, father of the President, stands on two-wheel wooden ox cart, talking to his Vermont neighbors as he bids the Caravan team a safe journey To the west coast (newsprint)

A group of Vermonters who voted for Lincoln helped start the Caravan on it's way. Left to right - Charles G. Wheeler, 80; Christopher Hall, 89; Leazer Hall, 87; R. E. Hathorne, 81; and Andrew Brown, 83. (newsprint)

E. A. Seidel (left), driver of the lead car for the Coolidge-Dawes Lincoln Tour, accepts the challenge with a handshake, prepared to begin the cross-country journey of a life time

WHERE IT ALL BEGAN – PLYMOUTH, VERMONT

Coolidge homestead in Plymouth, Vermont, where Calvin Coolidge took the Oath of Office for President in the middle of the night on Aug 3, 1923

The Coolidge Homestead is certainly the centerpiece of the life of Calvin Coolidge, involving a number of significant events in his life. He moved to the Coolidge home when he was only four.

It was here that, at the age of 12, he endured the death of his mother, Victoria Josephine Moor, in 1885. Calvin, his father Col. John Coolidge, and sister Abigail were at her bedside. In his autobiography Coolidge wrote, "The greatest grief that can come to a boy came to me. Life was never to seem the same again."

Perhaps the homestead is best known as the site where Calvin Coolidge, upon the death of Warren Harding, became President, his father, a notary public, performing the swearing-in ceremony in the homestead's parlor at 2:47 a.m. on August 3, 1923.

During their years at the White House, the Coolidge family continued to enjoy their summer visits to Plymouth Notch and the homestead. It was during these years that the homestead saw many dignitaries walk up the steps to pay their respects to President Coolidge.

The Coolidge Homestead is an important component of the activities of the Coolidge Memorial Foundation. It is also the centerpiece of the President Coolidge State Historic Site in Plymouth.

Plymouth Notch, the birthplace of a president, remains virtually the same today as it did a century or more ago. The homes of Calvin Coolidge's family and neighbors, the community church, cheese factory, one-room schoolhouse, and general store have been carefully preserved, and many have their original furnishings. The President is buried in the Plymouth Cemetery with seven generations of Coolidges.

It was here, in the center of Plymouth Notch, that the Coolidge-Dawes Lincoln Tour started its trek down toward New York City and then took the Lincoln Highway on to San Francisco, eventually involving more than 100,000 vehicles. Col. John Coolidge, the President's father, standing on a two-wheeled wooden ox-cart, welcomed his neighbors and friends and they bid the tour a safe and successful journey.

The Coolidges entertain at Plymouth Notch. Left to right: Harvey Firestone; President Calvin Coolidge, autographing a maple syrup barrel; Henry Ford; Thomas Edison; young John Coolidge standing; Mrs. Grace Coolidge; and Colonel John Coolidge, the President's father.

Frequently when the President and First Lady were visiting in Plymouth during Coolidge's administration, they entertained a steady stream of guests on weekends. Both Grace and Calvin Coolidge can be seen shaking hands with visitors in this photo. The President's father, John Coolidge, has his back to the camera as he observes the stream of people passing by.

The *Republican Champion*, Newport, New Hampshire, printed the following article in their September 11 edition:

Coolidge and Dawes Lincoln Tour Draws Quite a Crowd Despite a Downpour of Rain

The Coolidge Dawes Lincoln Tour, which started its trip across the country Tuesday afternoon at about 4:15 p.m. Instead of at 2:00 o'clock as expected.

The square had been filled with automobiles and people from early afternoon, during which time the Van Arnam's Minstrels Band rendered a concert in front of the town hall, after which the large American flag with the names of Coolidge & Dawes at each side was raised from the town hall to the new steel 80 ft., flag pole, which had been erected in front of Emerson Block.

The caravan, which consisted of seven automobiles, including a truck with the pictures of President Coolidge's birthplace on each side and some of his slogans printed thereon, was escorted into town by several automobiles from Claremont and several from this town.

Hon. George A. Fairbanks introduced the speakers in the order named from the town hall steps: Capt. John G. Winant of Concord republican candidate for Governor; U.S. Senator Henry W. Keyes of North Haverhill; U.S. Senator Simeon D. Fess of Ohio; and Congressman A. W. Jeffries of Nebraska, who spoke briefly but effectively in the interests of the campaign.

The caravan left for Keene in a downpour of rain after a visit of only one half hour here.

The tour was unavoidably delayed at the start in Plymouth, Vt., the birthplace of President Calvin Coolidge.

George E. Lewis was in charge of the affair locally and was ably assisted by other prominent republicans.

Congressman E. H. Wason of Nashua was with the caravan, but did not speak here. One noticeable feature of the afternoon was the fact that the members of Fred Smythe Post, No. 10, G.A.R. were present to greet the caravan at Claremont and to escort it here.

MAJOR "PLAYERS" IN THE COOLIDGE-DAWES LINCOLN TOUR

Those who traveled the entire length of the tour, or at least a good part of the distance, consisted of:

Tour Manager: John P. Cowan, Pittsburgh, Pennsylvania
Tour Secretary: Charles T. Doyle, Pittsburgh, Pennsylvania
Main Speakers:
Congressman A. W. Jefferis, Omaha, Nebraska
Senator Simeon D. Fess, Ada, Ohio
Herbert L. Moore, Plymouth, Vermont
Senator James Wadsworth, New York
Senator Theodore Burton, Ohio
Congressman Robert Simmons, Nebraska
(Dozens of other speakers participated along the route but did not travel for extended lengths of time with the Caravan. Many are mentioned in the day-to-day news coverage.)

Caravan Lead Vehicle Drivers:
E. A. Seidel, New York City, Reo Speed Wagon Caravan Lead Car
Ernest Hoch, New York City, Ford Amplifying Unit
S. E. Mollden, New York City, Dodge Press Car
"Bill" Wallace, Omaha, Nebraska, Dodge Speaker's Car
Ted Brandberg, Portland, Oregon, Dodge Speaker's Car
A. W. "Ab" Jefferis, Jr, Omaha, Nebraska, Studebaker Press Car
C. E. Thompson, New York City, Lincoln Speaker's Car
"Syd" Lewars, Newark, New Jersey, Lincoln Speaker's Car
G. A. Rickenbaugh, Cleveland, Ohio, Lincoln Speaker's Car
J. A. Smith, Chicago, Illinois, Lincoln Speaker's Car
L. L. Hasbrouck, Des Moines, Iowa, Lincoln Speaker's Car
Bill Pitton, Denver, Colorado, Lincoln Speaker's Car
C. L. Welti, Salt Lake City, Utah. Lincoln Speaker's Car
C.M. Hellbach, San Francisco, California, Lincoln Speaker's Car

(Note: Drivers of the Lincoln continued to switch as the Caravan moved across the country. It is believed these drivers were provided by various district offices of the Ford Motor Co.)

Press Representatives: A number of newspaper reports and columnists traveled for periods of time with the Caravan, although not for the duration of the trip.

Congressman A. W. Jefferis, director of the Caravan, and the man who nominated Charles Dawes for Vice President at the Cleveland convention

John M. McCullough of the Philadelphia Inquirer, covering the Caravan trip

E. A. Seidel poses by the lead vehicle

Caravan crew poses during stopover in Gettysburg. L. to R.: Pennsylvania Lincoln driver; A. W. Jefferis, Jr; Bill Wallace; S. E. Mollden; E. C. Hoch and E. A. Seidel.

Wednesday, September 10
Day 2

Left Greenfield, Massachusetts at 8:00 a.m.

Log: Stops at Northampton, Holyoke, Springfield, Westfield, Pittsfield, Mt. Lebanon, Massachusetts and Troy and Albany, New York. 200 miles. Spent night at Hotel Ten Eyck in Albany.

Review: First motor-cycle escort of the trip takes us from Springfield to Westfield at 50 miles an hour pace.

Added speakers today were Congressman Treadway, and Speaker Gillett.

New York State flag presented at Mt. Lebanon. Senator James Wadsworth joined us at Albany, and Senator Fess left us.

Parade consisting of 300 cars and a band of music escorted us through Troy and Albany.

E. C. Hoch, driving the Ford, meets with first accident of the trip, skidding into the rear of the Reo, and smashing radiator and headlights.

Pace much faster today, averaging about 37 miles per hour between stops.

—E. A. Seidel

At Northampton the half dozen motor cars comprising the caravan visited the modest Coolidge home where Mrs. Coolidge's mother, Mrs. Andrew Goodhue, now resides. Judge Henry B. Field, Coolidge's first employer in the field of law, was master of ceremonies and James Lucey, the president's cobbler friend, who yesterday was nominated for the legislature, also took part.

By 1:00, when the meeting had ended at Westfield, the caravan had an escort of over 100 automobiles for its journey to Pittsfield by way of Jacob's Ladder. Congressman Allen T. Treadway journeyed from Stockbridge to Greenfield early in the morning to accompany the caravan on its trip throughout the remainder of Massachusetts.

Coolidge-Dawes Club, Albany County, New York jugate button. 1"

The Ten Eyck Hotel, Albany, New York

REO Speed Wagon

The *Berkshire County Eagle* carried a lengthy account of the caravan's visit to Pittsfield under the banner, **"Coolidge-Dawes Caravan Is Given a Warm Reception; Big Delegations from Various Parts of Berkshire Take Part in Parade and Subsequent Meeting at City Hall Park."**

Pittsfield citizens extended yesterday afternoon a very cordial welcome to the Coolidge-Dawes caravan on its arrival here from Westfield, a welcome in which many other towns in Berkshire joined through the medium of delegations of prominent citizens who motored here to attend the ceremonies incident to the reception here for the cross-country "covered wagon" campaigners. Following a parade through North street, Maplewood avenue and First street, a meeting was held at city hall park, with addresses by various speakers, after which the caravan departed for New York state via Lebanon mountain, intending to go to Troy for a meeting last night and later to Albany.

The caravan left Westfield yesterday about 1:15 and shortly before 3 o'clock was met just this side of Lee by a delegation of 135 automobiles containing prominent citizens from north, central and south Berkshire, members of the official reception committee from the Pittsfield Republican Club, Coolidge Club and Republican City committee and the Republican Club band of 18 pieces. In the leading automobile was Thomas E. MacDonald, marshal of the day. Accompanied by this escort, and with the band playing "Welcome to Our City," the caravan entered Pittsfield and made an impressive spectacle as it wended its way over the line of march previously laid out.

At the park a sizeable crowd had gathered and before the speaking exercises began, many of the curious ones in it inspected the "covered wagon," or rather truck, bearing paintings on either side depicting the birthplace of President Coolidge at Plymouth, Vt. The primitive, rural surroundings of the homestead were referred to later on by some of the speakers as an encouragement to other humbly born boys to aspire to high things in life.

Photographer William Benedict took a picture of the truck and also a group picture of the members of the caravan assembled beside it.

While the preliminaries were being attended to, the band played and then the musicians vacated the truck and it was utilized as a speakers' stand Major Harris F. Hall, president of the Republican Club, opened the exercises. He said that Pittsfield was greatly honored by being among the first cities in New England to extend a welcome to the caravan on its trip across the continent and was duly appreciative of this fact. He then presented Congressman Allen T. Treadway as chairman of the meeting.

As the first speaker, Mr. Treadway introduced Thomas Prentice, night superintendent of the General Electric plant, who took a prominent part as a stump speaker in the Roosevelt presidential campaign. The latter acquitted himself in a very effective manner as usual and aroused considerable enthusiasm. He said that the democratic candidates at the democratic national convention in New York were nominated by a crowd of people that was exhausted with "The Sidewalks of New York." LaFollette was nominated because he was a disgruntled politician and because the ideals of his life had not materialized. Calvin Coolidge was nominated because the entire republican organization of the United States, represented in its general assembly, voted unanimously that he was the man. At the present time, he said, Mr. Coolidge was filling out the unexpired term of the late President Harding and in so doing was carrying out the latter's policies as the late Theodore Roosevelt under similar circumstances carried out President McKinley's policies. The country knew nothing of the Roosevelt administrative policies until after his election to office. Neither could it know anything of what Calvin Coolidge's real policies would be until he was elected the coming November.

The speaker then went on to liken "Silent Cal" to President Lincoln, the man of silence during the critical period of the Civil war and to General Grant, the sphinx at Gettysburg when the destinies of the nation hung upon this pivotal battle of the great struggle. He felt assured that President Coolidge would demonstrate his ability if elected, as these other silent men had done. He felt that the message that the caravan should carry across the country was the necessity of electing this able leader and to see that with him into office went the impetuous damn of General Dawes.

Congressman Treadway called attention at this point to the picture on the truck of the Coolidge homestead and stressed what an inspiration it ought to be to other boys in humble circumstances to go out into life and achieve great things.

Senator Simeon D. Fess of Ohio was then introduced. In opening, he called attention to a cane he carried which he had cut from the Coolidge homestead farm in Plymouth. He would treasure this cane all his life, he said, as an embodiment of the sentiment that was going to stir the hearts of the American people so long as opportunity is open for a man to walk out of a humble place like the homestead of President Coolidge to head the greatest nation known in the history of the world. That was the chief glory of America. That was the finest tribute that could be paid to any nation on earth. The open door swinging open on both hinges to any boy, whether born in a humble hut or not. For this reason he felt that the nation was going to do what Maine did Tuesday and what Massachusetts did and elect Coolidge and Dawes and the other republicans on the ticket. If the country believes in genuine Americanism, it should not hesitate as to what it will do with President Coolidge. If it believed in character; if it thought soundly; if it believed in economic recovery out of the conditions of war; if it believed that a way ought to be provided for the investment of every American dollar in American business; if it believed that every man who wanted to work ought to be able to find the opportunity to be employed at good wages in American industry run by American employers; if it believed in the maintenance of American business through the protective tariff that goes with the policy of the administration that Calvin Coolidge represents, then it ought to elect him. By the side of the president is one of the most militant, original and accomplished business men that America has yet produced, General Charles G. Dawes.

The speaker then went on to tell of the many high positions of national and international trust and responsibility that Mr. Dawes had been called upon to fill and how as the head of the reparations commission he formulated the plan that is now being adopted to stabilize Europe and thereby restore world peace and harmony. He was being accepted as a leader out of war disorders by the old world and America ought to be able to accept his leadership too. He trusted that it would.

Mayor Charles W. Power, who was late in arriving, was introduced at this point and extended a very brief, formal word of welcome on behalf of the city.

The concluding speaker was Cong. Albert W. Jeffries of Nebraska. He is a very massively built man of the aggressive western type and made a very favorable impression. In reply to a remark of Congressman Treadway's when the latter introduced him and referred to their comparative sizes, Mr. Jeffries said he had to grow tall, for he was raised between two corn rows and had to grow up to reach the sunlight.

The speaker paid much attention to General Dawes' abilities as evidenced by his part in the reparations discussions in Europe and how in accepting the Dawes plan, Europe had virtually accepted his leadership. This plan, so far as can be seen, is the only way out of the turmoil and hatred and economic prostration of the old world and if followed out faithfully will restore harmonious, brotherly relations among the peoples of the world and lead to that spirit of co-operation which alone can bring desired results, as no man or nation can stand along today and do his or its best. In its passage across the continent, he said the caravan would carry the message that it was vital to elect Coolidge and Dawes in November to ensure that America would continue to lead in this new era of world co-operation and regeneration. He also felt that such other party representatives as Speaker Gillett and Congressman Treadway should be sent back to the nation's capital where their services will be needed to back up the administration.

With a few concluding words of farewell by Congressman Treadway, the meeting adjourned with the band playing "Hail, Hail the Gang's All Here." The procession then reformed and the trip to New York state was begun with an escort of 35 automobiles and the band accompanying the caravan to the top of Lebanon mountain. On the mountain, Congressman Treadway formally turned it over to a delegation from Troy. A flag of New York State was presented to the caravan by the president of the Women's League of New York and was accepted by Congressman Jeffries.

Among the out of town men noted yesterday at the meeting at the park were Mayor Greer, ex-Senator George H. Hastings and C. Q. Richmond of North Adams, Postmaster James Young of Adams, ex-Senator John B. Hull of Great Barrington, County Commissioner Robert Tillotson of Lenox and Reuben Pierce of Dalton."

The New York Times: **Special Report**

**Albany Greets Caravan.
Coolidge Auto Tour Attracts
Crowds Everywhere.**

Albany, NY, Sept 10 – Albany and Troy Republicans turned out in force tonight to welcome to the State the caravan which left Plymouth, Vt, home town of President Coolidge, yesterday, to travel to the Pacific Coast.

The caravan, consisting of seven automobiles filled with speakers, met with enthusiastic receptions at every town and hamlet in the territory through which it journeyed today. Escorted out of Pittsfield by several hundred automobiles, the caravan crossed the New York line at New Lebanon and was met by an Albany reception committee headed by Mayor Clinton, with hundreds of automobiles. A New York State flag was presented by Mrs. Frederick D. Colson, President of the Coolidge and Dawes Republican Club of Albany.

The caravan was met here by a brass band and escorted to Odd Fellows Hall, where Senators James W. Wadsworth Jr, and Simeon D. Fess of Ohio made speeches. Former Representative Albert W. Jeffries of Nebraska, in charge of the trip, told the audience of its purpose. Senator Wadsworth warned his audience of the possibility of governor Bryan being the next President, if the decision was thrown into the House of Representatives. He urged the Republicans to get out and vote on Election Day.

Judging from today's events the new form of advertising a national election is a success. Everywhere people lined the roads along the route and at scheduled stopping places they swarmed about the speakers and cheered for Coolidge and Dawes."

THE CAMPAIGN OF 1924

The presidential campaign of 1924 in some ways was an enigma from start to finish. There were interesting, even fascinating "quirks" about the campaign, while at the same time being almost totally lacking of color.

Calvin Coolidge as the Republican candidate was the incumbent president, yet not well known by the public. As a Vice President under Warren G. Harding, he made very few contributions and there was a strong chance the Republicans would drop him from the ticket in 1924, but Harding's sudden death made Coolidge the new President on August 2, 1923.

Actually, Coolidge was an ideal candidate for the GOP. The Teapot Dome scandals of the Harding administration were never connected to Coolidge and he was able to continue the Harding Administration's program of "normalcy" and gain the confidence of both the American people and the Republican leadership, becoming one of the most popular men to occupy the White House.

Coolidge was quickly nominated on the first ballot at the Republican Convention in Cleveland, and Charles G. Dawes, a former Chicago banker & highly decorated general during the World War was selected as Coolidge's running mate. The convention was so dull that Will Rogers suggested the city open up the churches to liven things up a bit.

The Democratic side was a different story. Meeting in Madison Square Garden in New York City, John W. Davis, a former West Virginia Congressman and a Wall Street lawyer, was a compromise candidate at the Democratic convention. Davis won the nomination after 17 days in the middle of a heat wave...on the 103rd ballot. The split in the Democratic convention was huge, with a deadlock between pro-Ku Klux Klan delegates backing Californian William G. McAdoo (Woodrow Wilson's son-in-law) and anti-Klan delegates solidly behind New Yorker Alfred E. Smith. The fact that Smith was a Catholic was a particularly tough issue. Charles Bryan, Governor of Nebraska, whose main qualification was that of being William Jennings Bryan's brother, was the Democratic vice presidential candidate.

In 1924 the Progressives, unhappy with the selections of both of the major parties, decided to form a party of their own. They met in Cleveland for their convention which was attended by 1,200 delegates and 9,000 spectators. They launched a strong team as their standard bearers – Progressive Republican Senator Robert LaFollette of Wisconsin and Democratic Senator Burton K. Wheeler of Montana. The Progressives carried a strong message aimed at a coalition of farmers and workers.

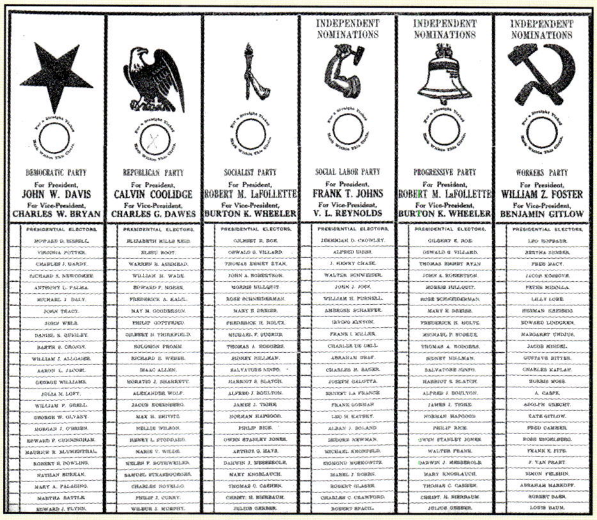

Official 1924 Presidential Ballot showing the candidates and names of the electors for the parties on the ballot

LaFollette and Wheeler also became the standard bearers of the Socialist Party in 1924, appearing on the ballot both for the Progressives and the Socialists in a some states. The Communist Party (or Workers Party), running William Z. Foster and Benjamin Gitlow for their candidates, called the Progressive Party platform "the most reactionary document of the year."

The 1924 campaign was both traditional and non-traditional. While Harding had been famous for his front porch campaigning, Coolidge did very little campaigning at all, leaving most of the traveling and speech-making up to his vice presidential candidate. The two major campaign thrusts included the activities of the Home Town Coolidge Club and the 56-day Coolidge-Dawes Lincoln Tour, a coast-to-coast political auto rally involving over 100,000 vehicles following, for most part, the route of the Lincoln Highway. Coolidge was not personally involved in either activity.

Interestingly, on the Lincoln Tour, with hundreds of speeches being made, Davis's name seldom was uttered. The Democratic candidate was totally ignored. The major concern was the campaign of the Progressives and the fear that a large LaFollette vote would throw the election into the Congress whereby, somehow, Charles Bryan could end up with the presidency!

Meanwhile, John W. Davis traveled the country, in state after state, perhaps racking up more miles than any candidates before him, but with little media notice.

On election day, it was a landslide for Coolidge and Dawes. They won nearly 16 million popular votes (382 electoral) to Davis and Bryan's eight million (136 electoral). Although the GOP's fear of LaFollette did not come to pass, the Progressives had a very strong third party showing of nearly five million popular votes and 13 electoral votes (Wisconsin). The Progressives ran second to Coolidge in several western states, including California.

Supreme Court Justice Oliver Wendell Holmes, Jr. summed up the Coolidge landslide this way: "Why tempt fate by opposing him? While I don't expect anything very astonishing from it, I don't want anything astonishing." This, indeed, captured the country's mood–Coolidge and normalcy was what the people wanted.

Election of 1924 Candidates	Party	Electoral Vote	Popular Vote
J.Calvin Coolidge (Mass.) Charles G. Dawes (Illinois)	Republican	382	15,725,016
John W. Davis (West Virginia) Charles W. Bryan (Nebraska)	Democratic	136	8,386,503
Robert M. La Follette (Wisconsin) Burton K. Wheeler (Montana)	Progressive	13	4,822,856
Herman P. Faris (Missouri) Marie C. Brehm (California)	Prohibition	0	57,551
Frank T. Johns (Oregon) Verne L. Reynolds (Maryland)	Socialist Labor	0	38,958
William Z. Foster (Illinois) Benjamin Gitlow (New York)	Workers' (Communist)	0	33,361
Gilbert O. Nations (Wash., D.C.) Charles H. Randall (California)	American	0	23,867
W.J. Wallace (New Jersey) J.C. Lincoln (Ohio)	Commonwealth Land (Single Tax)	0	2,778

1924 Presidential Election Results by Political Party

1924 Presidential Electorial Votes by State

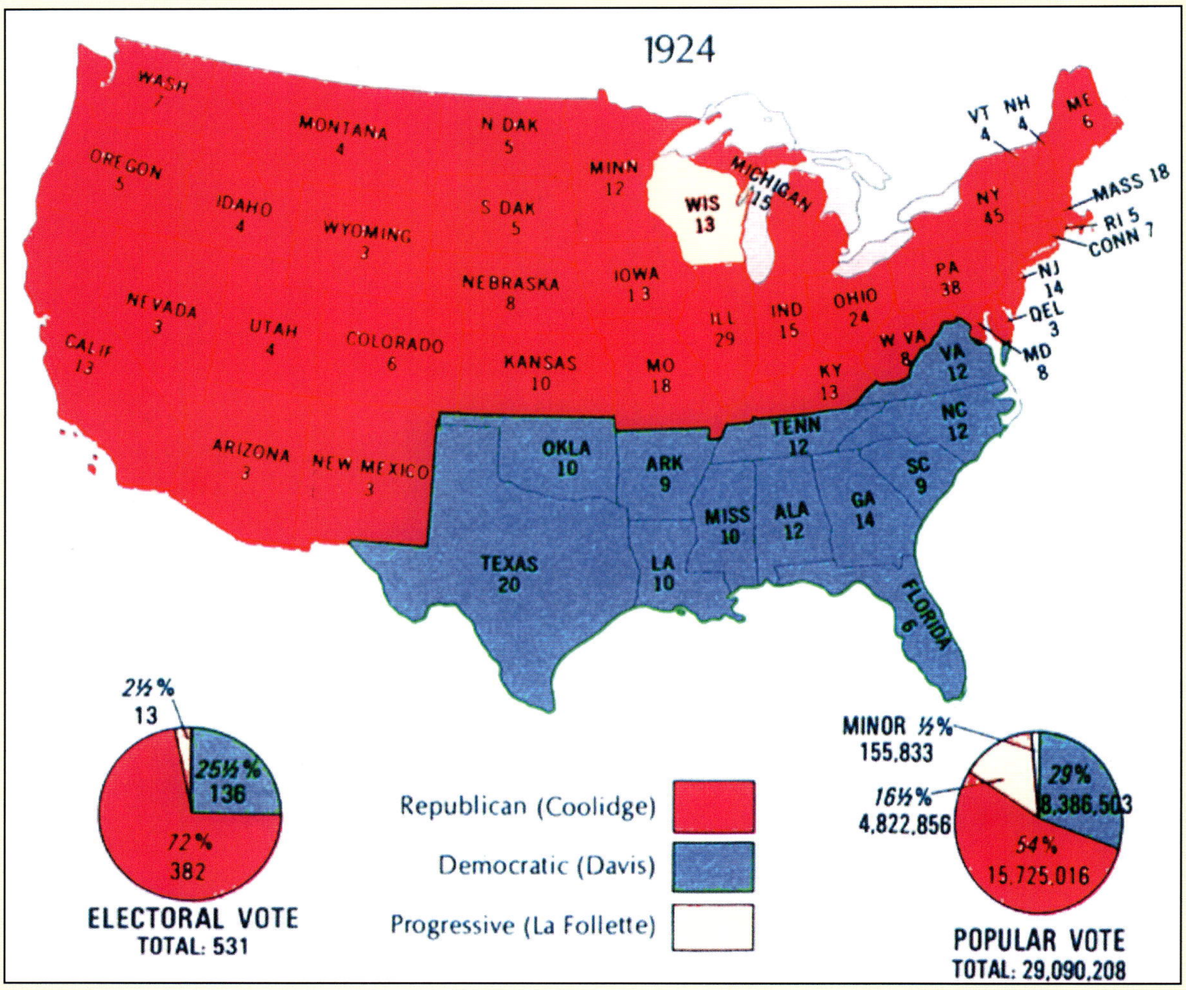

Thursday, September 11
Day 3

Left Albany, New York at 9:00 a.m.

Log: Stops at Schenectady, Amsterdam, Little Falls, Herkimer, Illion and Utica, New York. 90 miles. Spent night at Hotel Utica.

Review: H. L. Moore spoke for the first time at Illion, N.Y. and with his "Down East talk was a riot." He sure went over big, with his story about the boyhood days of Calvin Coolidge.

Weather fine today, and large enthusiastic audiences greeted us everywhere.

Senator Wadsworth spoke at all the meetings, and left no doubt as to his popularity.

The Ford truck could not stand the fast pace and was left behind at Albany.

Ab Jefferies, driving the Studebaker, had two flats just outside of Little Falls.

Bill Wallace, driving the Dodge, skidded into the rear of the Reo smashing his left front fender.

Dinner given to us at Illion, New York by Republican Club. Senator Wadsworth leaving us at Utica to speak at Syracuse State Fair.

—E. A. Seidel

The day following the visit, September 12, the *Schenectady Gazette,* carried a brief article on page 21 of their paper:

Republican Caravan Passes Through Here

The Coolidge-Dawes Lincoln Highway Tour caravan paid a brief visit to this city yesterday morning when it passed through here on its way to the West. A brief Republican political meeting was held at Erie boulevard and the party continued on its way to Amsterdam and other cities in the Mohawk valley.

The cars forming the caravan party were met at Elbert street by several hundred cars from this city and were escorted to Erie boulevard by Mayor William S. Campbell in his own machine. The Second Regiment band formed a part of the escort.

Senator James W. Wadsworth and Congressman A. W. Jefferies of Omaha rode on front of the truck which was decorated with pictures of the Republican candidates.

In Erie boulevard Mayor Campbell introduced the speaker, Senator Wadsworth who gave a brief campaign address. Senator Wadsworth apologized for not being able to talk longer and spoke but a few minutes. The speech consisted mainly in enrolling the personalities of the two Republican presidential nominees and the government of the present administration. He also emphasized the need of electing one of the presidential candidates, either Coolidge or Davis, in order to keep Governor Bryan from occupying the presidential chair, which might happen if the election of a president went to Congress.

Pair of scare 4" celluloid easel buttons

> It is interesting that although the *Schenectady Gazette* did not apparently cover the caravan rally on September 11, the day of the visit, nor announce the plans thereof, they did, in the first section of the paper, run a biting satire on the caravan direct from Washington, DC:

Over the Plains to the Oil Fields
For G. O. P. Caravan

Assistance of Forbes, Fall and Daugherty Needed to Make It Historically Correct

WASHINGTON, Sept 10 — The motorcar expedition of Coolidge- Dawes speakers to the Pacific Coast calls to memory earlier expeditions into the open spaces for the "winning of the west" for Republican candidates, notably the speaking tour of Charles E. Hughes in California in 1916, as well as the tour of the "Golden Special" beyond the Mississippi in the same year. It will be recalled that the tour of the "Golden" or "millionaire special" came to grief before it had penetrated far beyond the Mississippi, and that these and other efforts expended beyond the Father of Waters won the electoral vote of four western states for the Hughes-Fairbanks ticket, Iowa, Minnesota, South Dakota and Oregon. The "Golden Special" was a deluxe train, upon which all reservations were held by wives of Republican millionaires, heiresses and members of the "Four Hundred." Success of this tour was on a par with Hughes' speaking campaign in California, so "cool" did the men and women of the west manage to keep upon the occasion of the coming of the "millionaire deluxe special."

But this is 1924, the age of the motor car, and the 1924 caravan will be strictly an "oil-burner."

"Over the Plains to the Oil Fields" is suggested as a fitting title for the 1924 pilgrimage into the wide-open spaces of the west. Publicity agents for the Coolidge-Dawes caravan have broadcast the news that the tour will cover 5,000 miles and that it is likely 200,000 automobiles will be used. Such a train of cars is likely to draw heavily even on the administration's oil reserves.

There is unusual interest and inclination to join in this unprecedented cross-country demonstration and "trucks bearing floats will be in line, while farm tractors and horse-drawn vehicles will have places in the official escort," says the announcement of the trip.

That there will be interest in the pageant is undoubtedly true, as much or more for what it does not portray and for where it does not go as for what it does show and where it does go. But it is respectfully suggested that horse-drawn vehicles have no place in the caravan – there should be none but oil-burning vehicles in line. It is also mentioned that among the floats should be at least one representing a teapot. If the pageant is to be historically correct – and "common sense" will admit of nothing short of that – the caravan's route will be cleared by a squad of Boston policemen, while the escort will be furnished by "minute men" – not of the 1776 type, but of the Dawes 20th century variety. Also Colonel Roosevelt and a squad of Marines should be attached as part of the escort for in winding its way through Teapot Dome, a few troublesome "squatters" may be encountered. Some of Attorney General Daugherty's most vigilant sleuths might be employed with good effort, too – no well organized army would think of penetrating the enemy's country without a company of scouts – for there are United States Senators in the states the caravan will traverse and bootleggers are still aboard in the land.

According to former Representative Frederick C. Hicks, stage manager for the caravan- comedy, leaving Plymouth, the birthplace of President Coolidge in Vermont, September 9, it "will reach its goal, the Pacific coast, October 18," but if the caravan maintains no greater rate of speed than the President flashed in convicting Fall and Sinclair, or than he showed in ousting Denby and Daugherty, it is extremely doubtful that California will be reached in October of this year or next, particularly if it visits all those points of interest – in the light of recent events – that it will have to visit if it adheres to the historical truth of the present administration's record notably the "Little Green House on K Street," in Washington, Washington Court House in Ohio; Meadows of Dan, in Virginia; Teapot Dome, in Wyoming; Three Rivers in New Mexico, and Elk Hills, in California.

Northampton, Mass, the present home of President Coolidge, is down for a stop. Northampton is also the site of a veteran's hospital, so of course Colonel Charles R. Forbes will be on hand there to extend a hearty welcome to the pilgrims, while Harry Daugherty is expected to attach himself to the parade at "the Little Green House on K Street" and continue as far as Washington Court House. Bascom Slemp will be expected to head the reception committee at Meadows of Dan and in Wyoming, where the teapot will be sizzling, there will be a social hour. In true western style, Albert B. Fall will no doubt arrange a "rodeo" at Three Rivers, while E. L. Doheny will do whatever social honors the occasion requires, as well as fill the motor tanks with a fresh supply of oil and gas from his well known storage tanks.

It is understood that one of the feature attractions of the performance to be staged at each stop will be a moving picture presentation of the Dempsey-Carpentier prize fight, which film has heretofore been shown only in Washington.

The following personalized report on the caravan's visit to Little Falls is from the *Little Falls Evening Times* under the heading "**First Republican Rally of Campaign**."

Winding across the country, much in the same manner as many years ago, the famous log cabin of "Tippicanoe and Tyler Too," with its little brown whiskey jug, was journeyed overland by horse drawn vehicle the Coolidge-Dawes motor caravan stopped briefly in this city yesterday afternoon.

And there were several venerable local citizens, who had cast their ballots for the great emancipator, Lincoln, on hand to greet the party and to sign their names to the register which is to be presented to President Coolidge at the end of the transcontinental journey.

There was Charles F. Bowman, civil war veteran, old-time harness maker; William Dale, master builder; George Stebbins, whose pen still chronicles history, as it made history in voting for "Honest Abe," Albert Davy, well-known retired farmer of Davy's Corners, town of Danube; Thomas Wood, veteran of the Civil war, who served under Lincoln; "Doc. C. L. Isham, youngest of them all and radiating vitality and possessing a splendid physique which has been the result of long hikes over the hills and frequent draughts of cold spring water that gushes from the rocks at Lovers' Leap; and Whiiler Knapp of Dolgeville, sturdy lumberman.

Little Falls is proud of having so many of these splendid citizens living here. It was not expected that so many of them were still living, and they are to be congratulated upon their splendid and sturdy physique that has enabled them to remain to witness so great a change as has taken place in the country since their youth.

Local citizens began to gather in Ward Square at 2:30, the hour originally set for the arrival of the caravan. The warm sunshine, after a long and tedious spell of rainy and chilly weather, was conducive to good nature and they took the long wait, from that hour until the cars arrived at nearly four o'clock, with fine spirit.

A motorist passing through eastward inquired from a local citizen if the world fliers were going to stop here; another wanted to know what was "the rip" and if there was a fire, a murder or a big funeral. He was informed that such was not the case, and proceeded with alacrity. The genial chief "Dusty" Long was on hand to keep the passing autos moving and he did; take it from those who saw him do it.

Someone started a real flurry of excitement among some of the waiting folk when they made the statement that the president himself was going to come here with the caravan, and the excitement on the edge of the group was heightened when someone said: "That's Cal speaking now." Every neck in the crowd stretched a bit longer then retreated like a snapping turtle's head, when another bystander said: "Aw, shucks, that's nobody but Jim Wadsworth."

Altogether it was a merry party by the time the cars arrived. County Chairman Theodore Rogers briefly apologized for the late arrival of the caravan in this city, but "Ted" failed to state the true reason for the delay, which was given a Times representative by Senator Wadsworth, who laughingly remarked that the party had heard of the fame of the street fairs held in Fort Plain and had stopped there and visited the midway – of course building a few political fences around it while they were there.

Congressman H. P. Snyder, who acted as chairman of the local meeting, briefly introduced the speakers, after remaking that he was pleased to see so many of the school children here, because they are soon going to grow up into "good Republicans."

Former Congressman A. W. Jeffries of Omaha, Nebraska, who nominated General Dawes for the vice presidential berth on the national G.O.P. ticket at Cleveland, was the initial speaker, saying in part:

"Women and Men, Boys and Girls: I have not brought this big stick here (indicating a huge cane which he carried) to have any bad effect upon you, but because it was cut in the old pasture, on the old farm at Plymouth, Vermont, where our present president and our next president, Calvin Coolidge, was born, where he was reared, where was moulded his splendid character, where he received, in the common schools, his early education and environment which has resulted in the making of this great American. So you see, boys and girls, I cherish this stick and I am going to carry it across the continent. I also bring a flag – the flag of a great state, which was presented to this caravan on the top of Mount Lebanon by the woman who has honored with the chairmanship of the women voters of the state. A great flag of a great state.

"We do not want the man who was nominated from Nebraska for vice president. We of Nebraska don't want to palm him off on you easterners as a man for vice president. The day after Charlie was nominated, the day after the Democratic convention had gone through all its long period of gyrations, I met a good woman, a Democrat woman, on the street in my home town, and she said to me: 'Jeff, who's the joke on, and what's the joke?' And we have since been wondering whether it was on John W. Davis or on W. J. Bryan. And she went on to add: 'You know what the joke is, don't you – it's Charlie.'

"You know what W. J. B. said about John W. Davis before Charlie was nominated, don't you? You know how he had poisoned all the west by his statement that no attorney for J. Pierrepont Morgan should be nominated. So we Nebraskans have got the idea that after Davis was nominated, the Democrats, to keep Willie's tongue quiet – you know he seldom talks now, unless someone pays $1 or 75 cents to hear him – they let him have Charlie for the vice presidential candidate. Now who's the joke on: Willie, or Johnny?

"So I say, this caravan, as it moves on, carrying the flag of the great state of New York, will carry with it the memory of this scene – of this intense loyalty and devotion to government, as led by Calvin Coolidge and exemplified by Charles G. Dawes."

Senator James Wadsworth was then introduced and spoke in part as follows:

"A visit to Little Falls is always pleasant to me and this is not the first. I always avail myself of the chance, because one of the best friends I have, one of your people, with whom I have served nine years in the congress – I wish he could serve nine more there – Homer Snyder lives here.

"Regardless of your political affiliations, let me say, I am sure you will agree with me that what we need most in government is character among our public officials. That element is exemplified in your representative in congress and it is a regret for me to know that he has felt himself compelled to retire and give some attention to his business. Maybe he will get back after a while – after he has had a little rest.

"It may be interesting to you to know that this idea of this caravan was conceived by the people of Plymouth, Vermont, not to impress you with the importance of Plymouth; not to display anything grandly or elaborately; but the idea in sending a message to the whole country, telling of the origin of Calvin Coolidge and the atmosphere in which he lived. And one word describes the whole idea perfectly – simplicity.

"Plymouth wants you to understand this man and that's why this caravan is crossing the continent."

The senator then launched into a vigorous attack upon those whom he characterized as political hypocrits, who are not frank and honest in their utterances, but who frame them with the idea of hoodwinking the people. "The day of fakir in politics should come to an end here and now!"

The speaker dwelt at some length upon the tariff phase of the campaign, drawing some excellent word pictures of the way in which tariff and tax juggling affect the costs of living to the great body of the common people and showing where and how the Republican administration had cut the tax burden. He then concluded his remarks with a stirring plea for the return of Coolidge to the White House, sending with him General Dawes, symboling the combination one of the most efficient teams that have ever been elected heads of this nation.

From this city the caravan continued the journey westward, being met here by a delegation of Herkimerites.

The *Utica Daily Press*, *Utica Observer-Dispatch*, and the *Utica Saturday Globe* all gave the Coolidge caravan extensive coverage. From the Press:

CARAVAN HALTS HERE AFTER DAY'S TRIP THROUGH VALLEY; WINS CHEERS ALL WAY FROM ALBANY TO UTICA,

Night Rally Is Marked With Enthusiasm of Old-time Gathering – Rousing Meetings at Fort Plain, Little Falls, Herkimer and Ilion"

Swinging its way westward on a jaunt to the Pacific coast by automobile, the Coolidge-Dawes Lincoln caravan rested overnight in Utica and will today resume its cross-country stumping for the Republican standard bearers.

Election of Calvin Coolidge and Charles G. Dawes was strongly advocated by United States Senator James W. Wadsworth, Jr., as approving the administration's thrift in vigorous addresses Thursday all through the Mohawk Valley.

Other speakers emphasized simplicity and direct contact as keynotes of the campaign.

UTICA GREETS VISITORS AT ILION

With the enthusiasm of an old-time political rally, the caravan arrived in this city about 7 p.m. escorted by scores of automobiles containing prominent Utica Republicans, men and women who had gone to Ilion to greet the visitors. Red fire was burning its brightest, and a band was playing lively airs when the caravan came to town.

Fatigued a bit from their long overland trip, which started at Plymouth, Vt., the home of President Coolidge, those in the modern "covered wagon" expedition were glad to stop over night and get rested for resumption of the long and tedious trip today. Crowds cheered them all along the route, and although the party was nearly two hours late in arriving at Mohawk Valley towns, everybody seemed to display great patience.

Six automobiles were in the party making the trip from end of the country to the other, but it is estimated a million cars will fall into the procession in escorting the speakers from one town to another along the route. An amplifying device was left behind in Albany as the truck carrying it could not keep up the terrific pace of traveling 40 to 50 miles an hour between stops.

Those who will go all the way through the 19 states and participate at the open-air meetings of the caravan are A. W. Jefferis, Omaha, Neb., former congressman; Charles T. Doyle, Pittsburgh, secretary; John P. Cowan, Pittsburgh; Charles E. Thompson, New York, eastern sales manager of the Lincoln division of Ford Motor Car Company; E. A. Seidel, William Wallace, A. W. Jefferis, jr, William Molidon and E. C. Hoch.

Senator Wadsworth left the caravan in this city to keep a speaking engagement for Defense Day at Fulton. State Senator Frederick M. Davenport, Clinton, who joined the party at Ilion, made a stirring address at the local open air meeting.

When William Ross Lee opened the Utica gathering, every bit of space in the old canal bed between Genesee and Hotel streets was jammed with humanity. Lacking a platform, similar to those erected in some of the other cites and villages, the speakers here mounted the rear of the caravan's truck.

Painted on either side of the truck was a picture of President Coolidge's humble home in Vermont and the words, "*Get Down To Brass Tacks*," were inscribed above the driver's seat.

Senator Wadsworth, the first speaker, emphasized the substantial reduction in taxation by the present administration and at the same time warned against the confusion and demoralizing effect upon the Nation were President Coolidge to fail to obtain a majority of the electoral votes, thus throwing the election decision in the hands of the present Congress.

He also urged voters to shake off the lethargy of other years and go to the polls and vote on election day to insure majority rule.

If you will remember the tremendous importance of the dollars and cents side of the government, said Senator Wadsworth, you can not escape the fact that Coolidge and Dawes merit your support when you go to the polls on election day. The thrift of this administration, the care of this administration, must merit your support.

If you will recall that through the efforts of the present Republican administration there was an assumption of industrial prosperity because of the reduction in taxes, you must give your support. For had not taxes been reduced there would not have been possible a resumption of prosperity, an end to unemployment.

And understand, if you will, that it was this administration which found utter chaos in governmental affairs out of which order was brought under the leadership of the late President Harding and Secretary Mellon.

Speaking of the possibilities of the election being left to the decision of Congress, Senator Wadsworth warned:

Remember there seems to be some elements that would rejoice in the confusion, would rejoice to have such a demoralizing situation overtake the United States.

Senator Wadsworth said he felt, however, there was little danger of such a situation developing.

My own best judgment is that President Coolidge will get a majority in the electoral college.

But if President Coolidge is not elected either by your votes in the ballot boxes or by the House, then in my judgment the man with the next best chance to be president is not Mr. Davis, the Democratic candidate, but Mr. Bryan, the candidate for vice president on the Democratic ticket.

Previously in his speech, Senator Wadsworth had explained that under the Constitution if no one of the candidates for president obtained a majority in the electoral college, the House of Representatives is empowered to elect a president, each state delegation casting one vote, a majority vote is required. Senator Wadsworth pointed out that if the present political affiliations in the House were adhered to in the balloting for the election of a president, it might be impossible for any of the presidential candidates to obtain a majority of the 48 votes which the House has under such a set of circumstances. In this deadlock, he explained, the Constitution provides that the man selected for vice president by the United States Senate would automatically become president on March 4 next.

Let me speak a word about the necessity of voting, resumed the senator.

Now some publicity has been given to the last national campaign. The League of Nations issue had been discussed for months in the newspapers, magazines and periodicals. Other issues had been presented to the public in various ways and means. Yet what happened?

Twenty-seven millions voted. Eighteen millions did not vote. That is not majority rule.

Senator Wadsworth said it was through a reduction effected by the Republican administration during the last four years that economic problems of the Nation had been solved.

Referring to Coolidge's silence, Senator Wadsworth said: Coolidge is thinking all the time. He has plenty of time to think because his tongue gives him the chance.

Less illustrious, perhaps, than William or Charles Bryan, but every bit an orator and convincing speaker was former Congressman A. W. Jeffries of Omaha, Neb. He did not regard either of the Bryans seriously and declared they should stay in Nebraska.

Mr. Jeffries carried a flag of the state of New York which had been given him by some Republican women at Albany as the days journey through the Mohawk Valley was commenced. The flag, he said, would be carried the rest of the trip and the people of other states would be told of the wonderful response to the caravan by the New York people.

Everything looks fine for a great Republican victory, Mr. Jeffries declared.

Although Fort Plain was not on the original schedule, the caravan was met at Nellieton by Republicans of Fort Plain and escorted to that village where a brief halt was made. Music by a band enlivened the occasion. Surrogate Fox Sponable introduced the speakers.

Delay In Reaching Little Falls

Despite the delay of an hour and a half in arriving into Little Falls, a good sized crowd was at Eastern Park, including many school children. Theodore L. Rogers, chairman of the Herkimer County committee, introduced Congressman Homer P. Snyder, who in turn presented Mr. Jefferis and Senator Wadsworth.

As in his other speeches, Senator Wadsworth emphasized the modest circumstances in the life of President Coolidge and how America was still the land of golden opportunity.

Screeching of sirens, blowing of whistles and tolling of bells announced to Herkimer folks the caravan had arrived about 4 p.m. President J. Fred Smith, Committeemen Charles Crandall, H. G. Munger and Col. John W. Vrooman met the visitors at the park in Little Falls and escorted them to Myers Park, Herkimer, where another open air meeting was staged on a pavilion.

Ilion's Greeting

Monument Square in Ilion was next visited by the procession of political pilgrims who were greeted by another large crowd. For the first time since they struck the Mohawk Valley, the caravan drivers distributed lithograph likenesses of President Coolidge and General Dawes, which were eagerly grabbed by outstretched arms.

Another high spot in the Ilion meeting was the talk by H. L. Moore, from Plymouth, Vt., the President's home town. He told how he had attended school with Calvin Coolidge, milked cows with him and done various chores about the farm. President Coolidge, he said, has always been fearless when it comes to honesty.

R. D. Worden, president of the village, presided at the Ilion meeting. Theodore Douglas Robinson joined the caravan at Little Falls and accompanied it to Ilion. No end of political candidates gathered at Ilion, where dinner was served at Osgood Hotel, and several cars with banners boosting candidates for other than national offices broke into the procession.

Voted for Lincoln

Men who had voted for President Lincoln were given an opportunity to sign a register carried by one member of the party. This book will later be presented to President Coolidge. Among those who signed the book along the Mohawk Valley were:

Little Falls – Charles F. Bowman, 82; William Dale, 83; George G. Stebbins, 82; Albert Davey, 80; Thomas Wood, 82; Joseph Earl, C. L. Isham, 82; Wheeler Knapp, 89.

Herkimer – Charles Gibson, 87; J. P. Davis, 86; Walter S. Serviss.

Ilion – Theodore Harter, William H. Doyee, 86, Jacob Hoeffler.

Utica – W. Lafayette Ames; George B. Fairhead, 90; George H. Klinck."

The biggest news coverage for the day came from the speeches by Senator Frederick M. Davenport: **"DAVIS NO LONGER FACTOR IN RACE; DEMOCRATIC CANDIDATE IS ELIMINATED AT THIS EARLY DATE, DECLARES DAVENPORT"**

Elimination of John W. Davis, Democratic nominee for president, was declared at this early date by Senator Frederick M. Davenport in his speech. The senator claims the national election now rests between the cautious progress of Calvin Coolidge and the swifter, experimental program of the LaFollette movement.

Neither the Davis platform nor the candidate fit into the present political controversy, except in a straddling fashion of compromise, states Senator Davenport.

The national election contest this year seems already to have narrowed down to a choice between Coolidge and LaFollette. Davis seems to be permanently out of the picture. Davis is a capable man, caught in an unreasonably difficult situation. His party machine was split in the national convention in the struggle with prejudice and sectionalism. Davis has grown into a genuine Wall Street conservative in recent years. His running mate, Bryan, is another type – a political extremist from the West. The Davis platform is a middle-of-the-road collection of the phrases and platitudes of the last political generation. Neither the platform nor the candidates fit into the present political controversy except in a straddling fashion of compromise.

Coolidge and LaFollette squarely face each other. LaFollette voices the discontent of the Nation. Now, discontent is something to be reckoned with. It is the business of parties and government to assist in allaying by wise means the discontent of the people. At least, it is the duty of parties to point out wise private or public policies to remedy and dissolve discontent.

Between Coolidge and LaFollette is a question of method, of plan. LaFollette proposes a heterogeneous program representing the points of view of the heterogeneous groups which compose his following. Some of his proposals are wise; more of them are doubtful, untried, uncertain. It is a movement of classes and groups and sections.

The Coolidge Appeal

The method of Coolidge is to appeal to national unity and the general welfare, to the simple and more fundamental and practical and tried policies of human progress. It is a difficult time in the history of mankind. It is a time for caution. It is a time for the taking to heart of the admonition of Kipling: The jungle is wide and the cub he is small; let him think and be still. This is the Coolidge way. He is himself a plain man from the Vermont hills. He has little dramatic quality or brilliant genius in the usual sense of the word. It is his simplicity, his integrity, his quietness, his faith, which appeal to the people. His countrymen recognize in him the simplest and the best that is in them. He is a different sort of a leader than God Almighty has at some other times raised up for them, but he is theirs. He is on their level, and within the range of their comprehension and their concern.

He proposes national thrift and economy. He lays great emphasis upon thrift, as Vermonters do and as most Americans in all the generations have to do either early in life or late. It is a timely insistence. Of course there are increases of function and enterprise which government must provide as a nation grows more enlightened and humane. But the American people know also that there is a growing tendency to extravagant expenditure and waste of public money, which needs the sort of check Coolidge proposes to employ. Not niggardly, but thriftily, is a good watchword for popular government."

The New York Times: **Special Report**

Utica, NY, Sept. 11 – Coolidge Dawes coast to coast caravan brought the national campaign into New York State today when it left Albany early this morning and toured half a dozen counties, and wound up with a large meeting here tonight.

Crowds turning out at Schenectady, Amsterdam, Fort Plain, Little Falls, Herkimer, Ilion and Utica to listen to Senator James W. Wadsworth Jr, former Representative Albert W. Jefferies of Nebraska, in charge of the tour, and local county Chairmen.

The caravan, consisting of several automobiles and a truck bearing a replica of the birthplace of President Coolidge at Plymouth, Vt, caught the imagination of the farmers particularly.

At least 10,000 people today heard the message of tax reduction and honesty in government. Apparently there were few Democrats in the crowds, for no one heckled the speakers.

Everywhere county chairmen promised a big turnout on election day. The Coolidge plurality, they said, would carry into office the entire State ticket no matter who was nominated. In many cases the Republican chairmen admitted that this would not be so easy if Governor Smith was renominated.

The only city where it was admitted that LaFollette would make any showing at all was Schenectady. Here County Chairman W. S. Hamlin admitted that the Wisconsin Senator would probably poll a large vote.

The only untoward incident occurred at Amsterdam Here the caravan drew up in front of a firehouse and the orators had scarcely got underway when someone, suspected to be an enthusiastic Democrat, turned in a false alarm.

At Herkimer the procession was met by Theodore Douglas Robinson and Representative Homer P. Snyder who escorted the caravan into Utica. The meeting here tonight was the largest of the day and the crowd cheered Senator Wadsworth's promise of a sane Administration under Coolidge and Mr. Jefferies' attach on Governor Charles W. Bryan."

Friday, September 12
Day 4

Left Utica, New York at 9:30 a.m.

Log: Stops at Cortland, Owego, Endicott and Binghamton, New York. 90 miles. Spent night at Hotel Arlington in Binghamton.

Review: Big receptions at Endicott and Binghamton. Red fire parades and large crowd lining the streets.
—E. A. Seidel

The ***Utica Observer-Dispatch*** covered the morning departure from Utica.

Under brighter skies than it has been favored with much of the way, the Coolidge-Dawes Lincoln caravan continued its tour this morning starting from the Hotel Utica with Cortland for its immediate destination. Acting as escort from this city was Robert H. Lamb of Hinckley, chairman of the Republican county committee, who was accompanied by Mrs. Lamb and his nephew Henry C. Lamb. The route taken was by way of Deansboro, Oriskany Falls, Cazenovia and De Ruyter. At Truxton, 10 miles beyond De Ruyter, Mr. Lamb turned his duties as escort over to Chairman Lord of the Cortland County Republican committee.

The reception given us has been wonderful, said Former Congressman Albert W. Jefferis of Omaha this morning. Everywhere we have been met by enthusiastic delegations and many people from towns not scheduled on our itinerary have come to ask us to make stops with them. Mr. Jefferis admitted that the daily series of open air speeches was a hard strain and he is one with physique and voice well calculated to stand it. He said that he hoped to induce United States Senator Wadsworth, who was compelled to leave the party today to keep an engagement to deliver a Defense Day address in Fulton, to accompany the party through the West.

September 12 was a national holiday – Defense Day – and the comments by the speakers did not allow the special occasion to go unnoticed.

The *Cortland Standard*, in their coverage under the heading **"Coolidge-Dawes Caravan Arrives,"** offered the following:

The Coolidge-Dawes Lincoln caravan arrived in Cortland at 1:45 o'clock, only three-quarters of an hour late on the scheduled run from Utica. There are five cars carrying the official party, a truck with an amplifying device that made the start from Plymouth, Vt. having been left behind at Albany as it could not stand the terrific pace of 40 to 50 miles an hour maintained most of the way between stops. Fifty-five cars were in line at Truxton to escort the visitors in, and others joined them at various points along the route. Along the way flags were flung from house fronts, school children and people on lawns in front of the farm houses waved greetings and cheered the tourists, and as the head of the procession approached Main street it was greeting by a salute of bombs and firecrackers.

The caravan came in town on Clinton avenue to Main street, south on Main to Fort Watson, to Church, and north to the old Normal campus, where a large throng had already gathered to hear the speakers.

The big open air rally was held at the new county building, the men, women and children assembling in good numbers in front of the structure. County Clerk Charles R. Lord opened the meeting with brief remarks, introducing R. E. Owens of Cortland, who briefly reviewed the record of the present administration, referred to the duties of the citizens at the polls and commented on the decreasing interest shown by thousands of voters of late years in exercising the rights granted to them at the ballot box.

Former Congressman A. W. Jefferis of Nebraska, who had the privilege of notifying Charles G. Dawes of his nomination, was the principal speaker. The Westerner, strong of stature and with a commanding voice, held the undivided interest and attention of his hearers as he recited what the Republican administration has accomplished, showing by figures the sums saved yearly to the taxpayers, and touching upon the significance and real meaning of Defense day, which today is being observed throughout the nation.

The tour across the continent is mainly to preach the doctrine of Americanism as so well exemplified in the lives of the Republican leaders, and daily from five to six meetings are held. The splendid receptions accorded the caravan along the trail is but significant, said the former congressman, of the interest which the people of this great land have for that government which today is being guarded by two men who have its interests at heart and who are all the time working for America.

This is Defense day all over the nation, not a time when people desire to raise arms and a standing army, but a day which brings all just appreciation of our duties and obligations, and to demonstrate what could be accomplished if a storm might arise.

There is no other country so great as ours and we are looked up to as leaders by all the other nations, it is a land formed from the blood of all civilization and all living under one flag.

We all live in the spirit of love and cooperation, live and let live as well as give and take. We have gone forward in all our occupations, industries and educational movements under noble leadership. Is this all the result of some accident or superhuman power, or rather the fine institution upon which this government was founded? It is strong fundamentals that made possible present conditions and the fact that we are really what we are today.

The speaker said that he would quickly make hand and voice against any of those who are preaching to change the form of government and constitution. If such a day should ever come, and God forbid, what would then become of the freedom of the people, the press, religion, the habeas corpus act, and other privileges citizens enjoy.

The speaker referred to the fact that President Coolidge was criticized frequently because he expressed himself so little on national matters. He is a man who thinks a whole lot and speaks but a little, he said, but when he speaks his words are logical and carry weight and conviction. Speaking of the president, the Nebraskan cited the service the executive has rendered in helping cut down the government expenses since taking the chair. Figures were read from a statement of the bureau of budgets showing how the big indebtedness had been reduced from twenty-six billion to twenty-one billion, and how the money saved to the people of recent years was fully three times that of foreign trade, the railroad incomes, the cotton trade and the wheat markets.

Reference was made to the foreign policies and of the 54 out of 58 policies enacted and approved by this government...

In closing, the speaker referred to the New York flag which had been presented to the party as they crossed from Massachusetts into the Empire state and which he carried in his hand. He expressed the hopes that the flags of all the different states might be on hand in the procession of the inaugural ceremonies when Calvin Coolidge and Charles G. Dawes took office on March 4, 1925.

The meeting concluded with an address by H. I. Moore of Plymouth...reference was made to the Home Town Coolidge Club organized there and which today has over 15,000 members throughout the land. The historical ceremonies of August last year were reviewed briefly as well as the recent vacation spent by President and Mrs. Coolidge at Plymouth. The president is of the type of man, stated the speaker, whom the people believe and trust in and especially the mothers of this country who feel that the homes are safe under his guidance. "Keep Cool and Keep Coolidge in Washington" formed the theme of the poem recited by Mr. Moore in closing.

The members of the party took luncheon at the Cortland House following the rally meeting, later departing for a meeting at Ithaca, thence Owego and at Binghamton this evening."

The following two features are from the two Owego newspapers – the *Owego Times*, a Republican paper, and the *Owego Gazette*, a Democratic newspaper – both giving a very different "spin" to the event.

First, the *Owego Times*:

COOLIDGE DAWES LINCOLN CARAVAN IN OWEGO FRIDAY.

Big Crowd Heard Speakers.

Congressman Jeffries of Omaha and Hon. L. H. Moore, of Plymouth, Vt., Gave Good Addresses; Caravan Cut Ithaca Out of Itinerary to Make Time to go to Endicott.

About 20 automobile loads of Owego people motored Friday afternoon to Willseyville to meet the Coolidge-Dawes Lincoln Caravan, supposed to be on its way from Ithaca to Owego, but on arrival at Willseyville the Owegoans and others, were informed that owing to the hour of the proposed Ithaca visit conflicting with the defense Day Program, that city was omitted from the Itinerary and the caravan was coming to this village via Dryden, Richford and Newark Valley, and as a matter of fact, the caravan reached Owego before the crowd returned from Willseyville.

Ray M. Colby, who had charge of the Willseyville trip, had telephoned State Committeeman Rodney Robinson, of Ithaca, to telephone Emory Mix, at Willseyville, at the time the caravan left Ithaca. When Mr. Robinson found that the caravan would not visit Ithaca, he notified Mr. Mix of the fact and upon the arrival of the Owego caravan in Willseyville, Mr. Colby was informed and was able to halt others who planned to arrive later.

A large crowd assembled on the park after 4:40 o'clock and a few minutes before 5 o'clock, Assemblyman Daniel P. Witter, of Berkshire, called the assemblage to order and expressed his pleasure, which he was sure was shared by the audience, in having an opportunity to hear former Congressman A. W. Jeffries, of Omaha, the man who nominated Charles G. Dawes for vice-president, and Hon. H. L. Moore, of Plymouth, Vt., a life-long intimate friend of President Coolidge.

Mr. Witter then introduced Congressman Jeffries, who was greeted with hearty applause.

The speaker said that he was charmed to speak to the women and men of Tioga County and he expressed pleasure to see so many of the former among his audience; glad to see that they grasp the situation and are ready to work. He said the people want a guarantee of happiness and free speech, and that the constitution guarantees these desirable conditions.

To listen to some, said Mr. Jeffries, one might think that the country had gone to the bow-wows. But such is not so. But do we appreciate our country? Do we appreciate the foresight and acts of our fathers in drawing the document which is the safeguard of our rights, which creates and defines the power of our courts, of Congress, that our liberty may never be taken from us?

We have one candidate this year who is such of five or six parties, on five or six platforms, with their crying drastic recommendations for changes in our government This man would undermine our Constitution. He would no longer allow the Supreme Court of the United States to be a check on Congress. What would this mean? It would mean the passing of our religious freedom, of our liberty; it would vitally affect every home owner, every man, woman and child; it would make Congress supreme.

With Congress supreme, the average man would be at the mercy of men who might be carried off by fanaticisms. It would mean the curtailment of free speech, of the freedom of the press; it would place all subject to an autocratic power of one branch of Government.

Is LaFollette trying to build up our country and its institutions, commerce, industry or is he trying to bring chaos?

Now where is the brotherhood of man so near full exemplification as in this Nation. Nowhere under one flag are there so many happy, contented and prosperous people as under the Stars and Stripes. Every blood, every creed contributed to put America where it is today, the greatest nation on this earth. Life is more real in this country than any other place.

Do you appreciate America? If so, we should appreciate its government. For 135 years the Constitution has been our protection and our safeguard. We can't afford to let people fool with it too much.

We are interested in the cost of government, too. You may not think that you contribute to the cost of Government, but you do, either directly or indirectly. The government has no money except that which it secures from the people through taxes. Ninety-nine per cent, of the people work, but one per cent clips coupons.

Who pays taxes? We all do, whether you know it or not. For everything bought, if it be a manufactured product at least, the ultimate consumer pays. Therefore all should be interested in the cost of operating the Federal Government.

President Harding and Congress enacted the budget law and Charles G. Dawes was called in to put the Government on a coordinating, business basis, which he did within a year's time. He made the Government of the United States a business organization.

Nominee Davis says that the Republican foreign policy has been a keyhole policy. Since Harding took office, 54 treaties have been signed with foreign countries, all looking toward peace and accord.

The speaker then dwelt on the part General Dawes played in the reparations agreement of recent date, pointing out that the great powers of the world have accepted it.

In commenting on General Dawes' methods, the speaker said that Gen. Dawes went in London to Downing Street to see Lloyd George and was told at the door that he "could see Mr. George the day after tomorrow." He asked where Mr. George was and was told and Gen. Dawes made his way into Mr. George's presence, and explained his errand. Mr. George asked him, Mr. Jeffries said, "Have you any authority?" and was told in reply: "If you have half as much authority as I have we can settle it quickly," and it was done.

Mr. Witter, at the close of Mr. Jeffries' address, introduced Mr. Moore, who was also greeted with applause.

He told the boyhood life of President Coolidge and that he was an active lad, who did not scruple to stay out of school when his father needed his assistance in farm work, such as potato planting and getting in the hay, ahead of a rain.

He said that Mr. Coolidge's record is a white line all the way from his boyhood to the White House, and is without one act to explain or for which to offer an apology. He spoke at length on Governor Coolidge's record on the Police strike situation in Boston, and said that politicians told him it was political suicide, if he adhered to his intentions, but he did stick to it and was re-elected Governor by a larger majority than he was given when first elected. Mr. Moore told of the information of the death of President Harding, being taken in a car six miles to the Coolidge home in Plymouth, at midnight, and that the information also requested the immediate taking the oath of office as President of the United States, and Vice-President Coolidge had his father, a Justice of the Peace, administer the oath in the presence of about six people, by the light of a kerosene lamp. Then President Coolidge retired and at 6 o'clock in the morning arose and left for Washington. He refused a special train and had his car attached to the regular train, and thus began his journey to the National Capital, to assume his duties as President, and history shows that these duties have been honestly, courageously and efficiently discharged.

The party left for Endicott about 5:45 o'clock.

During the stay, a group photograph was taken with the prominent Owegoans posing on the courthouse porch and steps and in front of the steps.

And now, the coverage of the Coolidge-Dawes visit to Owego by the *Owego Gazette*, the Democratic newspaper:

THE COOLIDGE-DAWES CARAVAN VISITS OWEGO

Instead of the Outfit Coming by the Way of Willseyville, Where it was to be Met by an Owego Delegation, a Change of Plans is Made at the Last Minute, So the Caravan Comes Straight from Cortland to Owego – A Small Crowd Listens to Two Speakers.

The Coolidge-Dawes Lincoln caravan touring from Plymouth, Vt., to the Pacific coast, reached Owego at about 4:30 o'clock, last Friday afternoon not unsung, but unwelcomed. A considerable number of motor-cars from this village went to Willseyville expecting that the caravan would come by that route from Ithaca. On account of the fact that Ithaca was celebrating defense day quite elaborately the caravan's schedule for a stop in that city was canceled. It was expected to drive through the city only. Another last minute change was made, and the caravan came directly to Owego from Cortland over the Richford-Owego route. Hence the Owego welcoming party did not meet the caravan on the border of this county. The change was learned of through a telephone message, and the Owego contingent hurried back to this village, reaching here in time for the speech-making to begin a little after 5 o'clock, the time set.

The caravan consisted of a big limousine in the lead a REO speedwagon fitted up with specially built body on the side of which was painted the picture of President Coolidge's birthplace, a farm house at Plymouth, N.H., which for campaign purposes is termed the "humble cottage in which he was born." Banners and buttons were distributed from the rear of the speed-wagon.

Uncle Tom Was Missing

Some critic in the audience, not of the "G.O.P." faith it must be confessed, said the speed-wagon outfit looked as if it belonged to an Uncle Tom's Cabin show parade. Now this is a rather harsh remark, and only our duty as a truthful reporter compels the repeating of it. Far be it from us to cast aspersions on the unpretentious circumstances under which the president began his earthly career. If the reporter had happened to have been born in the log cabin that stood next door to his home when the stork arrived with him, he would have entered politics with his first vote.

Another unkind remark about the truck was that it resembled Andy Gump's caboose, which he is using in his presidential campaign. The caricaturist, who decorated the Coolidge truck, appeared to have appropriated some of Cartoonist Sidney Smith's ideas originated for Andy Gump's campaign.

Congressman A. W. Jeffries, of Omaha, Neb., who was the first speaker was introduced by Hon. D. P. Witter, of Berkshire, who presided. Mr. Jeffries turned his heaviest artillery from the north porch of the court house on LaFollette and made only a light assault on the Democratic standard bearer, John W. Davis. He spent a considerable portion of his time extolling the record of Charles G. Dawes in arms and in peace.

L. H. Moore, of Plymouth, Vt., a friend of President Coolidge in his boyhood, related principally incidents of his association with the president in boyhood-a sort of political "bedtime story," it was pronounced by another unregenerated Democrat in the audience.

The crowd that listened to the speeches was not large. Some of the people in the motor-cars which surrounded the park did not leave their cars, and could not hear the speakers. Not a few Democratic were in the audience.

A snap shot at the tourists and local Republican leaders was taken while the party was lined up on the court house steps. The caravan left at 5:45 o'clock for Binghamton.

Saturday, September 13
Day 5

Left Binghamton, New York at 9:30 a.m.

Log: Stops at Afton, Bainbridge, Oneonta, Grand Gorge, Roxbury, Margaretsville and Kingston, New York. 190 miles. Spent night at Hotel Stuyvesant in Kingston.

Review: Luncheon served at the Elks Club at Oneonta. Rained all day, and roads in very poor condition.

Skidded down a long hill, and made two cars coming the other way hit the ditch.

Large crowds greet us everywhere, the rain don't seem to make any difference.

Mollden and I taking Dodge to go home over the weekend. Arrive New York 2:30 a.m..

Caravan beginning to run very smoothly now and all stops are being made according to schedule.

—E. A. Seidel

Stuyvesant Hotel, Kingston, New York

The *Oneonta Herald* gave full coverage to the caravan's visit in the September 18 issue under the bold heading: **"Coolidge Dawes Lincoln Highway Caravan Holds Spirited Meeting at Municipal Hall Saturday, Raps LaFollette Candidacy."**

The Coolidge-Dawes Lincoln tour, one unique in the history of American politics, conceived by the boyhood friends and neighbors of President Calvin Coolidge, to acquaint the country with the Calvin Coolidge they know, stopped in Oneonta Saturday at about noon and held a spirited meeting at the Municipal hall which was largely attended by representatives of the party from all sections of this vicinity. The party came to Oneonta from Binghamton and went on their way to New York city by way of Kingston, leaving New York on the Lincoln Highway Tuesday for San Francisco and Seattle."

The two major speakers at Oneonta were Herb Moore and Ex-Congressman Jefferies. Moore gave his usual speech on the description of the early days of Cal Coolidge and Jefferies spent most of his time explaining the LaFollette plan to carry but five or six states with the hope of deadlocking the established method of electing Presidents, thus throwing the choice of a President into the House of Representatives. Jefferies also played to the audience, much to their delight, with his famous self-conversation regarding the Democrats: "William Jennings Bryan, my former fellow citizen," related the speaker, "denounced John W. Davis before the Democratic National convention. But the convention refused to follow Bryan and nominated Davis. But Davis was no sooner nominated than he asked William J. what he could do to keep his tongue off him. Well, I can't take the vice presidency because I've run for president three times, but there's my brother Charlie. So Charlie was named. Bryan is no longer a power in Democratic politics, but he attempts to dictate to the party."

"With regard to Defense day, Charlie Bryan said it shouldn't be observed and forthwith Davis agreed with Charlie. I wonder who is the Democratic candidate, William, Charlie or Davis?" And, so Jefferies went, stirring the crowd with his biting remarks on how he saw the Democratic leadership making their campaign decisions. And, as usual, getting a gigantic ovation at the end of the dissertation.

Music at Oneonta was furnished by the Company G band. Republican leaders were present from the entire Oneonta countryside including Cooperstown, Richfield, Unadilla and Schenevus.

In a report in the *Roxbury Times* it was reported that the Coolidge-Dawes caravan left Oneonta about half past three. the caravan, conducted by about 10 cars from Delaware county came through the Catskills where, at Stamford they were met by seven more cars from Grand Gorge and Roxbury. The first stop was at Grand Gorge, where Senator Arthur F. Bouton introduced Herbert L. Moore, boyhood playmate of the President who is now a member of the Vermont legislature. George L. Andress and Abram Yanson signed the register of those who voted for Abraham Lincoln. The next stop was at Roxbury where Mr. Moore again made a brief speech. A third stop was at Margaretville. All three villages were decorated in honor of the occasion with flags, buntings and pictures of Coolidge and Dawes. Good crowds were waiting at each stop, careful attention given the speakers, and much interest and enthusiasm manifested. At each stop Coolidge-Dawes buttons, emblematic of the trip, were distributed as well as printed explanation of the idea of the long trip on which they have started.

A week before the visit the *Roxbury Times* printed in full the letter they had received about the caravan's visit from the Republican State Committee

Republican State Committee
43 West 39th Street
New York, Sept 3, 1924

The Roxbury Times
Roxbury, N.Y.

My Dear Editor:

The Republican National Committee are planning for a Coolidge-Dawes tour which will go through the Eastern part of this State, arriving at Rensselaier county, Wednesday, September 10th and proceed as far west as Ithaca, thence to Binghamton on the 11th, and on the 12th will leave Binghamton and go through Afton, Bainbridge, Sidney, and other points to Oneonta and from Oneonta will go to Kingston on the State Highway through Stamford, Grand Gorge, Roxbury, Margaretville, Fleischmanns, and other places on the route.

The outfit will consist of a sound amplifying truck, cars containing literature, buttons, etc., and stops will be made at the various points and speeches made by eminent speakers of National reputation.

The Chairman and leaders of the respective organizations in the different counties and as many others with automobiles as can, are invited to join and accompany the caravan as long as it remains in the county or as much farther as they may desire.

A register will be carried in which all those voters now living who voted for Abraham Lincoln may write their names for presentation to President Coolidge.

I hope you will give this matter full and complete publicity and do all you can to make it a success.

Yours truly,
Lafayette B. Gleason,
Secretary

The *Catskill Mountain News* reported:

"The Coolidge-Dawes tour arrived in Margaretville later than scheduled on Saturday but stopped and one of the speakers made an excellent address in front of the Pocantico Inn. Three men signed the register of those who voted for Lincoln. They were Avery Hoffman and David Sliter of Kelly Corners and E. Hall of Andes. Automobiles from Margaretville went to Roxbury to meet the politicians and escort them into town."

The New York Times: **Special Report**

Coolidge Caravan Camps at Kingston. Farmers, Waiting in Downpour, Promise Republican Plurality in Up-State Counties.

Moore Rouses Enthusiasm. President's Farm Companion Quotes Letter Farmer Wrote in Boston Police Strike.

Kingston, NY, Sept 13 – Completing its tour of up-State counties on its coast to coast trip from the Vermont birthplace of the President, the Coolidge-Dawes caravan reached here tonight. Well satisfied with the reception in the cities and towns, the caravan will rest over Sunday and continue down the Hudson on Monday to New York City.

Leaving Binghamton this morning the caravan's progress was one of triumph, farmers and townsfolk lining the roadsides and waiting in a downpour of rain to cheer the traveling speakers and hear former Representative Albert W. J. Jefferies of Nebraska, director of the tour and Herbert L. Moore, President Coolidge's boyhood farmer friend in the Vermont hills tell why he should be kept in the White House.

Everywhere the spellbinders received assurance that New York State would do its duty on election day and roll up a plurality for the Republican ticket greater than that received by President Harding four years ago.

Republican County chairmen and local leaders asserted they found little indication that Democrats in their districts were enthusiastic over the outlook for their party this year. Many doubted if the Democratic vote for the need of the ticket would be even normal. There are evidences of LaFollette strength here and there are industrial centres, and the belief is general that the LaFollette ticket will run second.

Interest, especially among the farmers, continued in the new-found speaker, Herbert Moore, who told all day in simple, homely language of the boyhood life of the President on his father's farm, anecdotes of their work together during school holidays, and later got a ready response from the farmers who, in increasing numbers, are flocking to the towns where meetings are held. Mr. Moore tells his anecdotes in such a natural way and with such evident faith in what he believes to be his duty in presenting the real Coolidge to the nation that the natives listen spellbound, and give a mighty cheer when he ends. He told a new Coolidge anecdote today.

Mr. Moore related that he had seen a letter which Calvin Coolidge, when Governor of Massachusetts, sent to his father during the Boston police strike. This letter, he said, was written after Mr. Coolidge had decided to take a hand in the settlement of the strike.

"I am about to do something," the speaker said the letter read, "which may wreck my whole future political career, but, father, I consider it my duty to do this and I am going to do it."

In describing the small settlement of Plymouth, Mr. Moore said that little of news value had ever come to it until one of its natives had been made President of the United States. In forty years previous to that, he stated, the greatest event was when one of the Wilders put a new porch on his house.

Journeying through Afton, a small farming community in Chenango County, the caravan stopped long enough for Mr. Moore to tell his story. When he had ended Charles Seely, a retired farmer of the locality, 94 years old and said to be the oldest inhabitant of the county, hobbled to the little truck, which carries a painted canvas of the Coolidge birthplace, and asked to be allowed to sign the book containing the names of Republicans who had voted for Lincoln.

"Much to my regret, I never voted for Lincoln," said the old man. "I have always been a Democrat and I voted for Stephen A. Douglas. This year I am going to vote for Calvin Coolidge."

He was allowed to sign the book. About fifty Lincoln voters have already put down their names as having voted for the Emancipation President.

The best reception so far accorded the caravan was staged by former Representative William H. Hill, Republican boss of Broome County. Greeted at Binghamton by a parade and bands when it arrived there last night, the welcome extended into today, when it started for Oneonta. The caravan was escorted out of the city by Republican officials and Mr. Jefferies and Mr. Moore were called on to make their talks all over again. Today the caravan traveled through Broome, Chenango, Delaware and Ulster Counties. At every stop there was an enthusiastic reception, ending with a parade and a large meeting at Kingston.

An elaborate program is being worked out by the Republican National Campaign Committee for the caravan. Efforts are being made to have it welcomed at Peekskill by Chauncey M. Depew when it goes through Westchester County on Monday.

When the caravan leaves New York City next Tuesday to continue on its way to the Pacific Coast it will follow the Lincoln Highway and the Lincoln character of the trip will be emphasized. The caravan will use only Lincoln cars, said to have been given by Henry Ford, the President's friend. The cars will be in charge of Charles Thompson, who left Plymouth with the party and who will continue to the Coast.

One subject which the managers of the tour are to decide when they reach New York, is that of speakers. Several prominent members of the party will be asked to contribute their services. Since the party has been in New York, the only prominent spellbinders in addition to Director Jefferies have been Senator James

THE CARAVAN'S LEAD VEHICLES AND THEIR MANUFACTURERS

REO Speed Wagon - fondly termed the "modern covered wagon" - leads the way to California

Although the Coolidge-Dawes Lincoln Tour caravan saw well over 100,000 vehicles participate along the route from coast-to-coast, only a very few vehicles went the entire distance. At best, most of the rest went across a single state.

The six vehicles which made up the caravan's "lead entourage" included a Lincoln Speaker's Car, a Dodge Speaker's Car, a Studebaker Press Car, a Dodge Press Car, a Ford Amplifying unit, and a Reo Speedwagon Caravan Lead Car.

The Reo had photos of Coolidge and Dawes, a picture of the Coolidge homestead and campaign slogans decorating the sides and back. This made the Reo the most photographed vehicle of the tour.

Herb L. Moore, boyhood friend of Calvin Coolidge, poses by Speed wagon

"Bill" Wallace, driver of the Lincoln, poses by his car

It is strongly felt that most of these vehicles were donated to the tour by the chief executives of the respective auto companies who were supporters of the Coolidge Administration, but there is little documentation confirming this as fact.

We know that Henry Ford was a strong Coolidge supporter and it is believed that he donated both the Ford and Lincoln vehicles. Henry and Edsel Ford had purchased the Lincoln Motor Company at a receivers sale in 1922. Henry Ford took little interest in the Lincoln, however his son, Edsel, saw to it that the high standards of excellence that accompanied the Lincoln line were rigidly maintained and under the Ford ownership the Lincoln soon became a strikingly handsome automobile. Henry Ford had given Coolidge a Lincoln Sedan in 1923 while he was still Vice President and it is easy to assume that both the Lincoln and the Ford in the caravan's fleet were also donated.

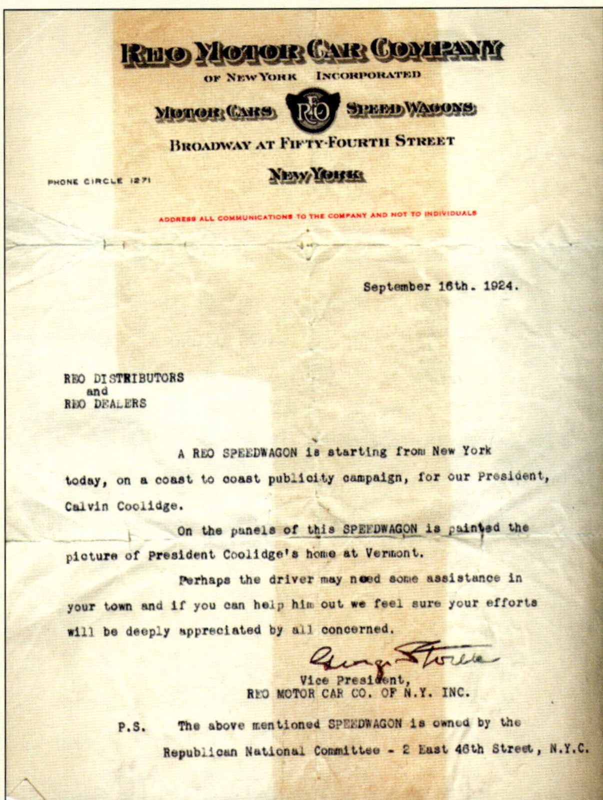

REO MOTOR CAR COMPANY
OF NEW YORK INCORPORATED
MOTOR CARS SPEED WAGONS
BROADWAY AT FIFTY-FOURTH STREET
NEW YORK

PHONE CIRCLE 1271

ADDRESS ALL COMMUNICATIONS TO THE COMPANY AND NOT TO INDIVIDUALS

September 16th. 1924.

REO DISTRIBUTORS
and
REO DEALERS

A REO SPEEDWAGON is starting from New York today, on a coast to coast publicity campaign, for our President, Calvin Coolidge.

On the panels of this SPEEDWAGON is painted the picture of President Coolidge's home at Vermont.

Perhaps the driver may need some assistance in your town and if you can help him out we feel sure your efforts will be deeply appreciated by all concerned.

Vice President,
REO MOTOR CAR CO. OF N.Y. INC.

P.S. The above mentioned SPEEDWAGON is owned by the Republican National Committee - 2 East 46th Street, N.Y.C.

A letter of correspondence from the Vice President of Reo Motor Car Co. to Reo distributors and Reo dealers across the country alerting them to the pending tour, and suggesting they provide any assistance to the driver of the Reo Speedwagon, if needed, also states that the Speedwagon is owned by the Republican National Committee. However, that does not necessarily mean that the GOP purchased the vehicle. Again, it may have been a gift, thus belonging to the Republican Party. Since the Reo Motor Car Company also ran display ads in newspapers along the route showing the decorated Reo Speedwagon, this could well be promotion in exchange for a gift of the van to the GOP. Ransom Eli Olds had founded the Reo Motor Car Company in 1904 after leaving the Olds Motor Works Company (Oldsmobile) which he had also founded. (When he left Oldsmobile he was precluded from using his name (Olds) for another auto company, so he chose to use his initials (REO).

It can only be surmised that both the Studebaker and the Dodge vehicles may have been donations or loans to the tour. Studebaker was doing very well during this era, in fact acquiring Pierce-Arrow in 1928. Likewise, Dodge was one of the major auto producers of the period with their one millionth Dodge Brothers car coming off the production line in the year 1924. Dodge Brothers was purchased by Chrysler in 1928.

The auto manufacturers may also have provided the drivers for at least a couple of these vehicles. We know the drivers of the Lincoln Speaker's Car changed as the Caravan moved along, involving at least eight different drivers coming from district offices of Ford Motor Company. E.A. Seidel, driver of the Reo Speed Wagon was the son of H. Seidel, sales manager of The Reo Motor Co.

We are not aware of any of the vehicles being driven from the final tour destination back to the East Coast. By the final day only the two Dodges and the Reo Speed Wagon were left. Seidel makes the notation that a local Reo dealer in Everett, Washington purchased the Reo, minus the body. It is not known what happened to this "work of art."

Sunday, September 14
Day 6

On Saturday evening, September 13, the *Boston Evening Transcript* carried a major feature, **"Caravaning The Coolidge Homestead Across the Continent"**. Authored by William A. Macdonald, this colorful feature of prose and descriptive narrative does an excellent job of reporting the first week's experiences of the campaign as a classic example of feature writing of the era:

The Covered Wagon, Painted with Portraits of Coolidge and Dawes and a Picture of the House in Which the President Was Born, Begins Its 5000-Mile Trip from Plymouth, Vt, to the Pacific.

Tonight and tomorrow, with six hundred miles behind it, the Coolidge-Dawes transcontinental caravan will rest beside the lordly Hudson at Kingston, N.Y., and on Monday the half dozen automobiles that compose it will wheel forward again on the rest of the 5000 miles that will take them to the Pacific, to San Francisco and then north along the western coast. They started on Tuesday from Plymouth, Vermont, from the home of the President's father. They rolled down through the rain into New Hampshire and on into Massachusetts on the first day. They left Greenfield under heavy skies on the second morning, crossing the state from north to south before swinging northwest from Springfield to ride through Westfield and Pittsfield and on into New York at New Lebanon. They swept on to Albany for the second night of the pilgrimage westward to Utica for the third night, southwest to Binghamton for the fourth and today moved back eastward on a long curving track to the great river whose course defines their pathway to Manhattan.

The caravan is the voice of the Republican party crying the names of its candidates in a land that so far has been no wilderness. It is the latest measure in campaigning, designed to reach the people in communities not ordinarily marked upon the itineraries of political speakers as well as the cities which are better known to older methods. It is the modernization of the campaigning formerly done by railroad extended to more than railroad comprehensiveness. It is new, conspicuous and picturesque. And, at the same time, it is as old as barnstorming – the old time political rally put upon a new footing of mobility.

The Original Idea

The idea upon which the tour is founded came from the mind of John P. Cowan, a Pittsburgh newspaper man, who marks its beginning in his mind as of last February. Originally the scheme was based on the cobweb: to converge a number of pilgrimages from different points upon a given centre. This proved upon full consideration to be impracticable and it was modified. Then it was suggested that a single caravan traverse the country from Maine to California. The proximity of Maine to the State in which the President was born provided the idea that the point of departure be the birthplace of the President and that was approved by the Coolidge Home Town Club, of which a founder, Herbert E. Moore, is one of the travelers to the Pacific coast. The Republican National Committee is back of the whole movement and Mr. Cowan is the manager now traveling with the caravan. Cowan, Moore, former Congressman Albert W. Jefferis of Nebraska and Mr. Jefferis's son are going all the way across.

Bostonians who remember the rain of last Wednesday here can visualize the storm in which the trip was started. The party which was due to assemble in Woodstock on Monday night was delayed at Rutland and in the night the rain began to fall. On Tuesday morning it whispered heavily in the trees and glistened on the roads as the motors wheeled toward Plymouth. As the morning aged the whole valley was sheeted in a curtain of water. Great tumbleweeds of mist that rolled in the mountains merged in solid vapor. The earth turned into mire and streams flowed underfoot. And so the caravan arrived at Plymouth.

At Plymouth

The group that was gathered there for the first meeting was a group of local people. They huddled on the expansive porch of the general store of the village of four houses. There were children from the school, given freedom for the occasion; there was half a dozen old men who had voted for Abraham Lincoln. The newspapers had their reporters there from Rutland and Boston and New York, and the cameras of the press clicked in the unending rustle of the rain. Here, at the window of an automobile, a veteran of the Civil War was questioned slowly as he told of voting for Lincoln as he turned from fighting; there on the porch of the store another wrote his name and age with uncertain hand, his body wrapped against the dampness and his feet encased in red rubber shoes. Senator Fess of Ohio conferred with Frank W. Agan, familiar to the politics of Vermont; and Colonel Coolidge, the father of the President, walked about the covered wagon which bears portraits of Coolidge and Dawes and pictures of the President's birthplace. He walked about the wagon, hat in hand, the rain testing on has bared head. The pictures seemed to interest him, but he walked there silently. If he though the pictures of the house a poor representation he did not say so. If he thought the portrait of his son looked too much like the portrait of General Dawes he did not say that either. Nor did he seem to mind the rain. Firm of step and observant of eye he plodded about the motored wagon which stood high above the other cars that made up the caravan. The son of a Lincoln voter had brought in an ox cart with its two beasts in yoke, and when Mr. Agan climbed upon this cart to make an announcement the colonel climbed up beside him. Nor was he less sure of foot than the younger man.

The meeting was held in the simple church across the street from the Coolidge homestead. Senator Fess told his audience gathered in the score of pews his high evaluation of Calvin Coolidge. Mr. Jefferis enunciated the first of the patriotic pronouncements that he will speak from coast to coast. The little crowd listened with appreciation that sometimes was audible in applause. And noon came and the pilgrimage was gathering itself to go. No power checked the rain. Colonel Coolidge sought the protection of the porch of the general store from which politics has stripped the remoteness that once it had. And as the group watched or gossiped the string of automobiles swung beneath the great elm before the store and swept out of sight around the curve in the road to Woodstock, the covered wagon near the van, and a trio of newspaper men, forgotten in the quickness of departure, asking Colonel Coolidge how to get a car to catch them up with the parade. With one word he found them a driver, with a single gesture he acknowledged their thanks, and suddenly Plymouth had seen the last of the latest invasion, and the caravan was splashing the mud of Vermont on its way to the mud of New Hampshire.

The Covered Wagon

Riding near the end of the group of cars, it was possible to mark its progress chiefly by the painted wagon that kept near the van. At every bend in the wooded road it towered its notification to the faces at the farmhouse windows. Here and there a woman waved or children called their excitement at the procession. And still it rained, but harder than before. Only by the end of the first thirty miles did it lessen, and then on their way down a long hill the tourists stopped, for there, at the bank of the Connecticut River on the line that divides the state of Vermont from the state of New Hampshire was drawn up a longer line of automobiles, and at the head of it was the national committeewoman of New Hampshire bearing the flag of her state, the first to be presented to the carrivan as it moved out of Vermont on its journey which will take it through seventeen states.

In the downcast day that party of reception had waited for several hours and among its numbers were figures familiar to the public affairs of the State and our day. Young Capt. John Winant, nominee of the Republicans in New Hampshire for governor, was one of these; a slim, keenfaced, dark young man wearing rather longish hair, speaking in a slow and quiet tone brightened by an occasional friendly smile. Later, in Claremont, and again in Keene, he addressed the crowds that gathered, accentuating his speeches with the gestures of a man not long acquainted with public speaking, and arousing responsive applause by the determination with which he said certain things.

The pilgrimage wound into Claremont, the line five times as long as when it started from Vermont. The rain, tired of reticence, set in again, but the crowd refused to be dispersed. In Newport, ten miles beyond, it was the same; the harder it rained the more people stopped to listen; and in this country where people are so Republican that little enthusiasm was expected, the response to the speakers was as warm as the rain was wet. The lights of the city glittered on the streets of Keene when the caravan had covered the more than thirty miles from Newport. In the crowd before the hotel were old men and young, men in working clothes, butchers from the markets, clerks from the stores, passersby of various station, girls in their teens, women in their fifties and sixties. A Chinaman accompanied by a Chinese boy drew a reporter down while he asked in soft liquids if Mr. Jefferis was the President. Is the President here? He persisted when his first question brought a negative. Twice, thrice he asked before he would be satisfied. Then, quickly, he turned, with the boy obediently in his wake, and left the crowd.

Oceans of Water

Nearly a hundred miles had passed beneath the wheels when the automobiles left Keene. They had been met on their approach by cars from Keene, but they departed with less ceremony. In the night the heavens dissolved themselves upon the world. In the shelter of a fortunate shed the driver of a truck replaced a tire. At Bernardston another car stopped to inquire its way. Headlights quivered upon oceans of water and the splash that sprang from the wheels of every car was as constant as the sound of the motor. Perhaps some of the pilgrims knew when they reached the Massachusetts line, but not all of them did. Suddenly the lamps of Greenfield, invisibly supported, set their row of glowing periods to the day.

Rain or no rain, it had been a successful day. Nine hours of travel and stops at three cities had brought out more interest than the officials of the party had expected. They talked it over with a satisfaction which modified any tendency to be tired. Wet clothing was no discouragement. The schedule had not been kept, but it could be kept, given better weather. This was proven when Wednesday began the journey through Massachusetts. It was a better day; it did not rain at all despite its heavy skies. The hills looked down upon the broad valley running south, the valley in which tobacco hangs greenly in the ventilated sheds. Through Deerfield, where the Indians once lifted the red axe for massacre, the campaigners rolled past prosperous farmhouses. They held their first meeting of the second day in Holyoke. They held their second meeting in Northampton. There was James Lucey, the cobbler, friend of the President. There were Mrs. Goodhue, and Congressman Treadway, and Speaker Gillett, smiling over his victory in the primary election of the day before. Newly informed with cheer the speeches went on in the town where Calvin Coolidge once made his home. The day before it had been the defeat of the Democrats in Maine that made a topical foil for optimism, but today the endorsement of the Speaker was an added subject for compliment. The meetings turned into cordiality for Mr. Gillett; there were little jokes, evidences of friendly warmth. The Speaker alluded to the successful beginning of the pilgrimage; the speakers of the caravan laid affectionate hands upon the Speaker's shoulder. The band played and the crowd applauded.

At Springfield the city committee had missed contact with the oncoming party and the meeting there was opened by Mr. Gillett himself in the square before the beautiful municipal group. It was a brief meeting, for the schedule had been adhered to and was not to be lost again. The sun gilded the noontime. With outriding motorcycle policemen sweeping ahead, the covered wagon betook its course across the bridge toward Westfield, its convoy in its train. Mr. Gillett remained with the party. Nowhere had he been received with such applause as he met at Westfield, his former home. Preceding him, Senator Fess grew emphatic in his elucidation of the consequences of an election thrown from the electoral college into Congress. He stated the difficult subject with simple clearness. Two conspicuous points of accentuation had already developed in the senator's talks: one the brilliance of the foreign policy of the Harding-Coolidge Administration, the other the socialistic danger of the third party. Mr. Jefferis balanced the arguments of the senator with an appeal to the sentiment, to the emotions of his hearers. He spoke of happiness in life and the need of preserving its opportunities by means of such a government as is in power. He spoke of the flag, of being American. Often introduced as the big man, with the accent on the adjective, he loomed above his fellows on every platform. Usually he gave his broad black hat into the hands of Senator Fess before he began to speak. And Mr. Fess, small, confident, smiling, in his turn had a trick of concluding his remarks with "Good bye" or "Much obliged" which seemed to please all the meetings with its unaccustomed informality.

Pittsfield's Welcome

It was Pittsfield that set up a new mark for welcome. It sent out 120 motors to meet the caravan. Down the road from Lenox, where the mountains empurpled the rim of the world with receding horizons, the long procession moved in a slow glitter of polished paint and twinkling metal. A bank upon a truck brazened the approach to the city. Speaker Gillett had dropped out, but Mr. Treadway still accompanied the journey. The flag of Massachusetts had already been added to the flags of Vermont and New Hampshire; the flag of New York was shortly to be companion to these. By four o'clock in the afternoon with a flare of sound from the motorcycles of the attending police and an escort of Pittsfield automobiles, the covered wagon took the road again and at the top of the hill where the States join was met by still more cars and more people and more speeches, while the chill wind bore again a brittle lance of rain, which perished almost before it could strike. And there, as the procession moved downward into New York, the hills unrolled from the broadest beauty of valley that had met the vision of the caravan since it left Vermont.

Mile upon mile spread its varied green. Trees were grouped like dark puffballs upon a field of emerald. Occasionally they were patterned in an orchard, again they massed their shadows as a wood. Farmhouses, white or dun or red, were separated by seeming miles. Church spires slenderly pointed from the centres of communities. Yet, as the train descended into New York, the houses by the roadside were of meaner quality than those of the New England towns that had been left behind. And, at the same time, the people seemed more interested or more curious than those across the border. Larger groups gathered in the smaller towns; more faces peered from the windows; more men and women stood upon the doorsteps. Once an aged negro, standing before a little white house splotched with the vivid colors of flowers, slowly waved his battered hat. At another place in a post of dignity before its household sat a fine gentlemanly black cat with spotless shirt-front against his sable coat, and even white cuffs. It seemed as if the house lacking a sign "Tourists Lodged" was an exception.

An Organized Campaign

It was not without careful organization that such a caravan had been enabled to find reception committees awaiting it. The work was done through the Republican National Committee and with the aid of State and county chairmen and other officials of the party. Now, as the party proceeds across the continent, it is preceded by its advance agents, one traveling by motor ten days ahead, the second about three days ahead. In the hands of State party officials it the itinerary of the trip, worked out even to the time of the various scheduled meetings. The only slip of the first two days was at Troy, which was quickly remedied. The demonstration there more than offset the delay. It was six o'clock in the evening when the caravan slid down the hill toward the red roofs of Albany. The golden outline of Hendrik Hudson's Half Moon moved uneasily upon the weathervane of the building beside the broad river. The speakers hurried still further north still to return for the night meeting in Albany. Every now and then a bomb announced the gathering of the faithful. The forces of Senator Fess and Mr. Jefferis were joined by the voice of Senator Wadsworth. And so from coast to coast new speakers will take the platform at new points. Through New Jersey and Pennsylvania, thence westward into Ohio and Indiana and Illinois and Iowa and Nebraska to Wyoming and Utah and Nevada and on to California and north to Washington new voices will speak for Coolidge and for Dawes. But always across the whole length of the Lincoln highway will go the tall bulk of the covered wagon painted with the portraits of the candidates and the pictures of the house in which the President was born. It will be the focal point of the attention of crowds that gather about whoever the speaker is."

Monday, September 15
Day 7

Left Kingston, New York at 8:00 a.m.

Log: Stops at Poughkeepsie, Peekskill, Tarrytown, Yonkers and New York City (149th Street, Bergen Avenue, 54th Street and Broadway). 120 miles. Spent night at Hotel Woodstock in New York City.

Review: Met by 500 cars at Hastings-on-the-Hudson, parade down the Hudson, through Yonkers, down the Grand Concourse, and through Central Park to Broadway and 54th Street.

Mollden smashed into the rear end of the Reo doing 45 miles an hour in Central park, pushing in his radiator and smashing both headlights and fenders. His brakes failed to hold.

Motorcycle police take us through Central Park and down Broadway 45 to 50 miles an hour, cutting Columbus Circle on the left.

This is the first truck that ever made a complete trip through Central Park at that rate of speed.

E. C. Hoch rejoined us at Kingston to act as relief driver.

Presented with silver "Victory" wreath at Yonkers.

—E. A. Seidel

The Peekskill local newspaper carried the following feature:

Coolidge-Dawes Auto Caravan Here Today. Prominent Republicans of This Section Greet Party North of Village and Escort It To Park Circle For Rally – Large Crowd Assembled. Nebraskan Promises Economy in Government. Congressman Jeffries Scores LaFollette – Declares Third Party Candidate and Davis Have No Chance of Election – Says New York Will Show Why.

The much-heralded Coolidge caravan reached Peekskill shortly after noon today. The party was met about two miles outside of the village by a committee of representative men of Peekskill and other towns of Westchester County. Among those who welcomed the Coolidge party were the following: Franklin Montross, James W. Husted, James K. Apgar, Congressman J. Mayhew Wainright, James Diamond, Milan E. Goodrich, and others.

Before a crowd of several hundred people who had gathered at Park Circle, Congressman Jeffries of Nebraska, the leader of the caravan, declared that the economy in government, for which Coolidge and Dawes stand, applies as much to the so-called poor man as to the rich man who makes out an Income Tax return. He illustrated this by showing how the costs are passed on from producer to wholesale merchant, retailer, and finally to the consumer.

Congressman Jeffries praised the budget system which Dawes installed, and pointed out how it had saved the nation the neat sum of two billion dollars.

He also called attention to the Dawes work in assisting in straightening out the European muddle. He said, in answer to charges that the present administration had done nothing in the matter of helping in settling post-war problems, that after Great Britain, France and Germany had wasted six years in trying to arrive at an understanding, Dawes was finally sent to Europe at the suggestion of the present administration. Mr. Jeffries stated that Dawes had gone over to Europe and talked "brass tacks' to the representatives there, telling them that they would have to sink their prejudices and animosities, and that they did.

Mr. Jeffries also scored LaFollette charging that the third party candidate had himself confessed that he did not expect to be elected, that his only object was to prevent the old parties from succeeding. The speaker pointed out that since neither LaFollette nor Davis had a chance for election, the thing left to the people was to elect Calvin Coolidge and Charles Dawes.

The speaker recalled how large groups of women from the different states through which the caravan had passed, had waited in the rain for hours to greet them and to present them with banners on behalf of the womanhood and motherhood of the states.

With the banner of the State of New York in his hand, Mr. Jeffries said that he would carry it through to the west coast and to victory.

The trip through Westchester down to the Bronx line is expected to be a triumphant procession. At Hastings which will be reached about 3 o'clock in the afternoon, 5,000 automobiles and trucks will form a parade and escort the party through Yonkers to 186th Street and Broadway, where the last meeting of the day will be held. No meeting has been scheduled for Manhattan.

Tuesday morning the caravan will start out on the Lincoln Highway for the trip across the country. There will be meetings at New Brunswick and Trenton, winding up Tuesday night in Philadelphia. Senator Porter H. Dale of Vermont will join the party in New Jersey for a few speeches.

In closing, Congressman Jeffries predicted that the State of New York would be carried by Coolidge and Dawes, and that the banner, which he held aloft, would be carried down Pennsylvania Avenue on the 4th of March next year at the inauguration ceremonies of the above-referred to candidates.

The Congressman spoke for over three-quarters of an hour, during which time he held the complete attention of his hearers.

The meeting was called to order by Franklin Montross of Peekskill, who introduced Honorable James W. Husted as chairman.

In his introductory remarks, Mr. Husted referred briefly to the home life of President Coolidge, and in a very happy manner, introduced as the first speaker, Herbert Moore of Plymouth, boyhood friend of the President, who was largely instrumental in the organization of the coast-to-coast caravan.

Mr. Moore is a very plainly dressed gentleman, unassuming in manner, and a person who gives the impression of being very democratic in his habits of life, if not in politics.

Mr. Moore spoke of the twenty-nine people who make up the President's village of Plymouth, spoke of the times when he had sat on a milking stool alongside of Calvin Coolidge and helped him milk the cows on the Coolidge farm, and explained as the reason there was no telephone in the home of the President's father at the time the news of ex-President Harding's death, that the old-maid housekeeper there would not answer phone calls. He closed by repeating a poem on "Keep Cool and Keep Coolidge."

Due to the lateness of the hour, the next speaker, Congressman J. Mayhew Wainright, spoke briefly. He stated that he was sure that the common sense and Americanism that had come out of Vermont would result in to the best interests of the country.

The meeting was adjourned about half-past one o'clock."

"Coolidge-Dawes Caravan Greeted by a Large Crowd; Much Enthusiasm When It Arrived Here Yesterday Afternoon," was the headline coverage in the *Tarrytown Daily News.*

Escorted by Motorcycle Officers Cregier, Dolen and Martin of the Tarrytown police, and Officer Taxter of the Greenburgh police, the Coolidge-Dawes caravan arrived in Tarrytown yesterday with banners streaming amid the welcoming strains of a lively air played by the Chevrolet band and the raucous roar of many automobile horns.

Chief Williams Bowles of the Tarrytown police and Captain McQullian of the Greenburgh police were on hand in their official capacities. All the officers not otherwise on duty about the town were stationed about the Washington Irving High School and at vantage points on Broadway, to keep traffic moving and handle the large crowd that gathered to hear the messages of the G.O.P. campaigners.

Supervisor Charles D. Millard was not on hand to give the travelers the official welcome of the town of Greenburgh, for, as he explained it later, he missed the caravan in Peekskill, where he went to meet it, and returned to Tarrytown after it had arrived. The official glad hand was therefore extended by Village President Ernest F. Griffin.

The Coolidge-Dawes caravan consisted of four big touring cars, and a special van fitted on a truck chassis. This was in the form of a boxlike housing, the four sides of which were covered with pictures of the candidates, of the Coolidge birthplace in Plymouth, Vt., and with appropriate slogans for the election campaign.

The touring cars belonging to the caravan were covered with blue and white pennants "For Coolidge and Dawes," and showed unmistakably the distance the cars have already traveled, this being, to the time they arrived here, over a thousand miles.

After welcoming the party President Griffin introduced as the first speaker Herb Moore, a Vermonter who worked side by side with Coolidge in the latter's farming days, and who knew the President, and the conditions under which he grew to manhood, as probably no one outside the Coolidge family does.

Moore told of Coolidge's birth and early manhood, described the village of Plymouth (which incidently has, in addition to four farmhouses, a store a church and a school, a population of exactly 29) and told with a ring of pride in his voice of Calvin Coolidge's rapid rise through ranks of political honor – from village trustee of Northampton, Mass, to president of the United States.

In conclusion, speaking of the President's policy of executive direction, Moore said, Our Great nation is too large a job for any one man alone. What we are looking for, we have exactly – a man who will sit down and counsel with his cabinet, decide on the best policies, and when the decision have been made, will stand squarely on his feet and fight for them.

In the meantime, Supervisor Millard had arrived from Peekskill, and he introduced the second speaker, Congressman Jefferis, of Nebraska, who is manager of the caravan tour...

Congressman Jefferis introduced a portion of the vast wealth of national history connected with Tarrytown when he said, Great history was made here, and I wish more of the vast number of people of this nation knew its lesson, when Williams, Van Wart and Paulding captured Major Andre, the spy whose work threatened to destroy all the careful work of the revolution.

A great crisis was reached when Andre very nearly made good his escape, and came very close to succeeding with his treacherous plan.

But then those three men whose names you have carved on the monument by the side of the road were there to step in when the crisis arose. They saw their duty and steadfastly they followed it. They feared neither the consequences, nor the possibility of failure. Theirs was the duty of soldiers – and nobly they performed it.

It is safe to say that had they not so surely reduced the plottings of the spy to naught, the development of this nation would have been retarded several hundred years. History would have been vastly changed. But they were there!

And just as there was, in the days of the revolution, a great crisis to be met, and brave men there to meet the need, as today there is a crisis, the resulting issue of which may also change the history of the nation.

There is a deserter from the ranks, that exponent and organizer of discontent, Robert LaFollette, who fundamental structure of the nation...

Col. J. M. Wainwright, our own representative in Congress, was next introduced. His remarks followed the trend of the previous speakers as he said, "Nothing is existing political history compares with the new force that threatens ominously, the whole structure, thought, progress and development of the nation as that which confronts the people of Westchester, and the nation today."

Colonel Wainwright stated in briefer form many of the facts told by the two previous speakers and concluded by saying, "We have never had a man who represented the Americanism of Abraham Lincoln, the homely virtue of the great man, better than Calvin Coolidge does today, and I am sure you will have no hesitancy about keeping in the President's chair next year, the man of the people and the man of the hour, Calvin Coolidge."9

The Yonkers Statesman for September 16 prominently covered the caravan's visit with news coverage and photos including the heading:

TERRACE CITY EXTENDS ROYAL GREETING TO G.O.P. CARAVAN

250 Yonkers Machines Meet Coast-To-Coast Visitors at City Line and Escort Party To Bronx

Amid the blaring of bands, the triumphant shriek of countless automobile horns and the welcoming cheers of thousands of onlookers, the famous Coolidge coast-to-coast caravan, outward bound from Plymouth, Vt., to Seattle, Wash., passed through

Young girl - a descendant of Charles Dickens - poses as Angel of Peace and presents wreath to Caravan officials

Yonkers shortly after 3 o'clock yesterday afternoon. The caravan was met at the north city line by 250 Yonkers machines loaded with city officials, Republican city and ward leaders, and Coolidge supporters. Its leaders declared the reception given them here one of best since they left Plymouth.

The Coolidge caravan left Plymouth, Vt., home of the Coolidges, last Tuesday morning on its historic campaign tour. It will follow the Lincoln Highway across the continent, being met in Pennsylvania by the President. In Chicago by General Dawes, and in California by the President again, according to present plans of the Republican National campaign managers.

There are six automobiles in the caravan, one of which is an automobile truck, showing pictures of the Coolidge homestead, portraits of the Republican National candidates, and hearing the slogan, "Coolidge and Dawes – Brass Tacks and Common Sense."

Members of the caravan party are: Herbert L. Moore of Plymouth, Vt, a neighbor of Colonel Coolidge, the President's father, and one of the organizers of the "Home Town Coolidge Club" of Plymouth, who delivers speeches daily wherever the caravan makes its stop; former Representative A. W. Jefferis, of Omaha, Neb., who hopes to "stump the country" from coast to coast; E. A. Seidel, driver of the Coolidge car, of the Bronx; Captain J. Huyler Ellison of New York City, former Secretary of Campaign Director Hicks; John P. Collin and Thos. Doyle of Pittsburgh, managers of the caravan.

The caravan reached the Hastings line on Warburton Avenue yesterday afternoon about 3 o'clock and found the 250 decorated automobiles of the Yonkers Coolidge boosters awaiting it. Its members were greeted by Acting Mayor Watson, Campaign Manager Robert M. Clark, Republican City Chairman J. Wade Hampton, and Campaign Secretary William D. Cameron.

Miss Dianne Brown, descendant of Charles Dickens and daughter of Mr. and Mrs. Walter F. Brown of Park Hill, dressed in the flowing white robes of "Victory," presented a silver victory wreath to the caravan. The wreath was attached to the front of the Coolidge car and will be borne to the Pacific coast.

Lincoln Voter in Party

Others who met the caravan at the North city line were August F. Schwarzler, Republican leader in the Bronx; Thomas W. Whittle, Surveyor of the Port of New York; James Hamilton, former Assistant State Treasurer; Representative A. B. Rossdale of the Bronx Peter Winn of the Bronx; and Henry Lerch of the Bronx, who voted for Abraham Lincoln for President, and who signed the Lincoln Voters Book in the Coolidge car, containing many names of men who voted for Lincoln and who will vote for Coolidge.

All the Yonkers cars made a gala showing in their decorations of banners, bunting, American flags, and campaign insignia. Three silver cups were awarded as the escorting fleet of machines passed in review before the judges committee on McLean Avenue. Miss Dianne "Victory" Brown, who rode in a coupe driven by Mrs. Robert M. Clark was adjudged the best "decoration" in the closed car division and she was presented with a silver cup.

Prize for Mrs. Hurst

Mrs. Albert Hurst of 683 Black Avenue won the cup in the open car division. With Mrs. Hurst rode Mr. James McPherson and Mrs. Howard K. Greenhaigh. James Ferrio was the driver of the truck that won the cup in that division.

The Route

The route taken by the procession was south on Warburton Avenue to Manor House Square, to North Broadway, to Getty Square, to South Broadway, to McLean Avenue, to Central Avenue, to the Grand Concourse in the Bronx.

Hundreds of cheering citizens lined the sidewalks on Warburton Avenue, Manor House Square, North Broadway, Getty Square and South Broadway as the caravan passed.

Many Yonkers cars wheeled out of line upon their arrival at the southern boundary of the city, but almost as many more accompanied the caravan through to the Bronx.

In the Bronx, a short speech was made by former Congressman Jefferis, who declared that the great question involved in the present campaign is whether or not the people will function in the government. The liberties of the people he said, are at stake in the candidacy of LaFollette for President, who would sweep out some of the parts of government which insures freedom. He urged his audience to be "true Americans" and to support a true American, Calvin Coolidge, for President.

Welcome in Bronx

Richard W. Lawrence, chairman of the Bronx County Republican Committee, welcomed the caravan in behalf of the Republicans of the Bronx.

A meeting was held at 149th Street, where the caravan ended its journey for the day. Here a crowd of several thousand cheered Thomas W. Whittle, Surveyor of the Port of New York, a Republican leader of the Bronx, as he introduced former Representative Jefferis and Herb Moore.

The caravan will make stops today at Elizabeth and Trenton in New Jersey and is expected to lay over for the night at Philadelphia. Several speakers, including Senator Porter Dale of Vermont, will help the caravan on its way through New Jersey and Pennsylvania. When it reaches Ohio it will be met by Senator Simeon Hess, who already has assisted it through Vermont and Massachusetts. United States Senator James W. Wadsworth, Jr., may also catch up with it again in Ohio.

A detachment of policemen under Captain Edward Connolly of the Fourth Precinct directed the traffic arrangements, and an escort of motor cycle police, under Sergeant Jerome Linehan, accompanied the caravan through the city."

Herb Moore telling it to the "mike" at Peekskill, New York

Major Sanders and wife, H. L. Moore and S. E. Mollden at Tarrytown, New York

Tuesday, September 16
Day 8

Left New York City on the Lincoln Highway at 8:20 a.m.

Log: Stops at Jersey City, Newark, New Brunswick, Princeton and Trenton, New Jersey. 85 miles. Spent night at Hotel Stacy-Trent in Trenton.

Review: Amplifying car supplied by Western Electric Co. at New Brunswick and Trenton.

Meeting at the Trenton Theatre, holding 3,000 people while 1,500 listen to amplifiers on the outside.

Party joined by ex-Governor Stokes, also Senator Edge.

The Hotel bill and breakfast is on the Jersey Committee.

Meetings all well attended, and motorcycle escort all the way.

—E. A. Seidel

The *Daily Home News*, New Brunswick, New Jersey, carried the following story:

Crowd of 300 Hears Neighbor of "Plain Cal" Coolidge Boost His Candidacy Here

The Coolidge-Dawes caravan rolled into New Brunswick this afternoon and with a crowd of 300 in attendance, staged a political meeting at the Pennsylvania Railroad Plaza. The caravan was met at Metuchen by State Committeeman Fred C. Schneider, county candidates and a large number of women active in the Republican ranks.

In the procession of automobiles, as they moved up Albany street, was a large float bearing a picture of a little farmhouse, the house of President Coolidge in the hills of Plymouth, Vermont. The automobiles in the procession had Coolidge-Dawes pennants suspended from the sides, and with the tooting of horns and klaxons, the caravan reached the plaza, where a large crowd had gathered.

The meeting was scheduled for 12 o'clock, but it was nearly 1 o'clock when the meeting got under way. State committeeman Schneider presided, and introduced the various speakers, who included Herbert L. Moore of Plymouth, Vermont, a neighbor of President Coolidge; Congressman Albert W. Jeffers of Nebraska, Congressman George N. Seger of Passaic county, United States Senator Walter E. Edge and ex-Congressman T. Frank Appleby.

Mr. Moore in his discourse on the life of President Coolidge told of many intimate details in connection with the President's life in the Green Mountain hills. He told of youthful days spent by Coolidge on the farm and of his education at college and his admission to the bar.

"He has always been plain Cal Coolidge to the people of Plymouth," said Mr. Moore. "He has never changed, and when he was sworn in as President of the United States it was in the simple manner, with his father administering the oath of office in their little cozy home, in which there is neither telephone or electric light," declared the speaker.

Congressman Jeffers directed his remarks to LaFollette, and declared that LaFollette was attempting to break down the Constitution of the United States and to set Congress up as an autocratic body to trample upon the weak.

United States Senator Edge welcomed the caravan to New Jersey, and declared that New Jersey was safe for Coolidge and Dawes. "The people of New Jersey will not stand for the radicalism of LaFollette, and true to traditions, the State will be found in the electoral college for Coolidge and Dawes on November 4," said Senator Edge.

Ex-Congressman Appleby added to the remarks of Senator Edge that he served and was associated in Congress with President Coolidge and enjoyed the association of Mr. Dawes while in Washington.

The last speaker was Congressman Seger, who spoke on the protective tariff and its relation to the industries and working people of New Jersey. He declared that the protective tariff and its relation to the industries and working people of New Jersey. He declared that the protective tariff was the salvation of the working people and the manufacturers, for without the high tariff many of the factories would be unable to compete with foreign manufacturers and it would result in hundreds being thrown out of employment here.

Following the meeting here, the caravan moved on to Kingston, where the Mercer County Republicans met it and escorted it to the Nassau Inn at Princeton for luncheon. Before coming to New Brunswick a big mass meeting was held in the high school at Metuchen, which was attended by 1,000 voters. The tour will be brought to a close tonight in New Jersey with a mass meeting at Trenton, and tomorrow the caravan will move into Pennsylvania.

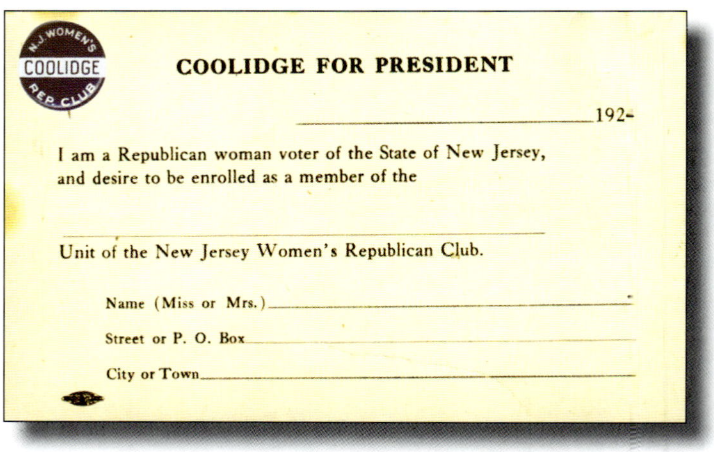
N.J. WOMEN'S COOLIDGE REP. CLUB

COOLIDGE FOR PRESIDENT

________________ 192-

I am a Republican woman voter of the State of New Jersey, and desire to be enrolled as a member of the

____

Unit of the New Jersey Women's Republican Club.

Name (Miss or Mrs.) ________________

Street or P. O. Box ________________

City or Town ________________

New Jersey Women's Republican Club membership card and button

The *Newark Evening News* for September 16 reports:

Coolidge-Dawes Party With Caravan Here Welcomed Noisily. Takes Course Through City Center with Elizabeth as Its Next Objective.

Bedecked in pictures of the birthplace of President Coolidge and likenesses of the President and his Vice-Presidential running mate on the Republican national ticket, Charles G. Dawes, a truck-load of literature escorted by a line of automobiles, passed through Newark today on its way from Plymouth, Vt., scene of Coolidge's birth, to the Pacific Coast over the Lincoln Highway.

The Coolidge-Dawes caravan, as the procession was officially termed by the Republican national campaign manager who arranged the tour, had a varying number of automobiles in line, ranging from twenty to thirty cars. As the line passed through the city behind a squad of motorcycle officers assigned by deputy chief of police to open up traffic automobiles dropped out and others joined the procession.

The caravan entered the city over the Passaic River-Plank road bridge, having come here from Weehawken, where the New Jersey state flag, to be carried on the journey, was presented by Senate President Firman M. Reeves. Its route then was west on Market street, to Springfield avenue, to High street, north to Central avenue, east to Broad street, south to Pointer street, west to Frelinghuysen avenue, south to Elizabeth city line.

Though horns were continuously blown in Market and in Broad streets and most of the cars in the line were bedecked with blue and white Coolidge campaign pennants and other decorations, few persons stopped on the sidewalks to see what was going by. The caravan line sped through the city streets without a halt until it reached the city line.

There, leaders of the tour posed for camera lenses while Newark's traffic police turned the situation over to a detachment of officers from Elizabeth. The caravan is to continue through Elizabeth, New Brunswick, and Trenton to Philadelphia where it is due to stop overnight.

Flags decorating B road street may have given some of the Coolidge caravan passengers an idea that the city was having a holiday to celebrate the caravan's arrival. True it is, the Republican Essex County Committee had announced that it would hold a meeting of Republicans at Proctor's Roof Theater tonight to celebrate the visit of the caravan. But the flags were out for another purpose-the opening of sessions of the National Waterways League Convention.

Among those accompanying the caravan through the city were Chief Prohibition Enforcement Agent George Zipf, Congressman Ernest Ackerman of Union County and Congressman George Seger of Passaic; United States District Attorney Walter. D. Van Riper and John P. Cowan and Charles T. Doyle of Pittsburgh, distributors of the Republican campaign literature.(Believe this is a Democratic newspaper.)

THE LINCOLN HIGHWAY AND ITS ASSOCIATION

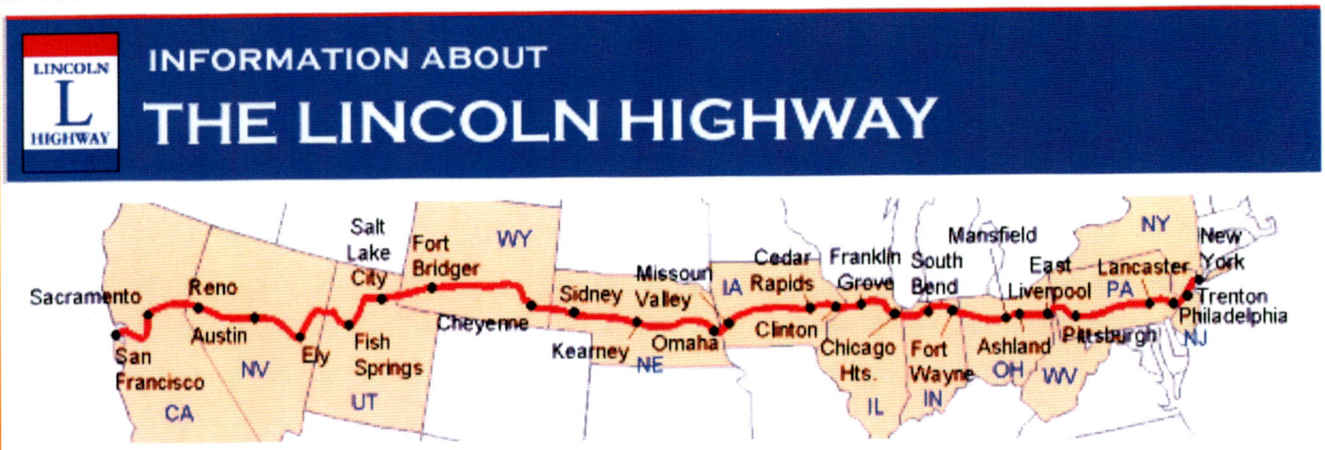

Route of the Lincoln Highway

The Lincoln Highway was the first continental automobile route in the United States, and the most famous road in America in its day. It was the catalyst of a "good roads" movement which ultimately involved the national government, creating the system of federal highways (in 1926) with a numbered state highway route system, and ultimately the interstate highway system (in 1956) that we take for granted today.

In 1912, most roads in America weren't suitable for the automobile. What few "improved" roads existed didn't go beyond the edges of towns and cities. Dirt roads led from one town to the next, and few people could give directions beyond that. There was no way to go any real distance, and certainly no way to drive coast to coast.

The idea of the Lincoln Highway came from the fertile mind of Carl Fisher, owner of the Prest-O-Lite Headlight Company and the man also responsible for the Indianapolis Motor Speedway and Miami Beach. Fisher was an early automobile enthusiast and believed that the automobile, as great as it was, wouldn't get anywhere until it had good roads to run on. Fisher called his plan The Coast-to-Coast Rock Highway, a graveled road that would cost about $10 million. He hoped Henry Ford would support the project and that auto manufacturers would contribute one percent of their income to fund it.

But Ford had other ideas. If private industry paid for roads, people would get used to that and expect it. He begged off. Fisher continued to promote his dream, holding a dinner meeting with many of his automobile industry friends on September 10, 1912 at the Deutsches Haus in Indianapolis. Eventually two other auto industry pioneers joined the campaign – Frank Seiberling, president of Goodyear and Henry Joy, president of Packard. A Lincoln Highway Association was formed, holding its inaugural meeting on July 1, 1913. Joy was elected president of the new association and it was his suggestion to dedicate the road to the martyred Abraham Lincoln.

Fisher, Seiberling, and Joy envisioned a hard-surfaced road that would stretch almost 3400 miles from coast to coast, New York to San Francisco, over the shortest practical route. The goal was to have the road completed by May 1, 1915, in time for the Panama-Pacific International Exposition in San Francisco.

Funds were raised through private donations from the public. All along the route, local associations were formed to raise funds and move the project forward. The great road would start at New York City's Times Square and go through New Jersey, Pennsylvania, Ohio, Indiana, Illinois, Iowa, Nebraska, Wyoming, Utah, Nevada and into California, culminating in San Francisco's Lincoln Park.

The route across America was determined by a number of factors, such as the need to connect major cities and supportive communities, including scenic landscapes and historic sites, and above all, follow the most direct route. Fisher describes the great Lincoln Highway in 1914 as "something more than a road. It will be a road with personality, a distinctive work of which Americans of future generations can point with pride – an economic but also artistic triumph."

Within a year of its beginnings, The Lincoln Highway Association was broke, but just beginning their crusade to enlist the public and the government to build better roads. The Association realized that its limited funds were better spent on promotion than on highway construction, especially after corporate America failed to provide strong sponsorship. They made the Lincoln Highway the centerpiece in a massive marketing campaign to convince the public that better roads were needed and that the government should build them.

Concrete "seedling miles" were built at locations strategically placed between notoriously muddy stretches of dirt road to serve as prototypes of what an improved highway could do for the nation. The idea of a major road crossing the nation was embraced by an enthusiastic public, and many other named roads across the country followed. The creation of the Lincoln Highway had a significant impact on how people traveled. No longer were they held to the schedules of railroads. Instead, more and more people chose to tour America by driving the Lincoln Highway, and as automobiling became more popular, the face of the roadside changed. Filling stations, tourist cabins, motor courts, restaurants and diners lined the Lincoln Highway to service travelers. A competition for the travelers' business increased, entrepreneurs became creative in their attempts to solicit customers, building restaurants and hotels in the shapes of boats, coffee pots and so on.

America's enthusiasm for good roads eventually led to the involvement of the federal government in building roads. By 1925, the web of roads had become confusingly marked and the American Association of Highway Officials instigated a policy of numbering all interstate roads, including all named highways. The Lincoln Highway that connected the country from coast-to-coast would become a series of disconnected numbers (U.S. 1, U.S. 30, U.S. 40, U.S. 50 and other designations).

The mission of the Lincoln Highway Association was completed. It disbanded on December 31, 1927 as an active

Carl G. Fisher

board after passing one final resolution to mark the route as a final memorial to Lincoln. The markers were designed by landscape architect Jens Jensen of Illinois, who won a national competition. They are cast concrete with a bronze head of Lincoln, the highway logo, and a blue directional arrow. In one day, September 1, 1928, Boy Scouts placed over 2,400 markers (weighing over 220 pounds each) at every mile across the coast-to-coast highway. Only a few of these markers remain today, signifying the route of the original Lincoln Highway – the greatest road in America.

Lincoln Highway Association

In 1992, nearly 65 years after the original Lincoln Highway Association disbanded, another group of dedicated historians met in Ogden, Iowa, for the purpose of preserving the memory of America's most historic road. They started a new Lincoln Highway Association.

Today, the Lincoln Highway Association is dedicated to preserving and promoting America's first transcontinental highway for the automobile.

Many stretches of the Lincoln Highway still exist in the street patterns of cities and towns along its route and as roads to nowhere on the edges of towns. Driving the Lincoln Highway, you will come across examples of commercial architecture from the 1920s and 1930s, bridges and other scenes, providing an opportunity to cross America and really see it, up close and personal.

The LHA works to not only preserve the Lincoln Highway, but to pursue appropriate measures to prevent further deterioration and destruction of the remaining sections of the highway. Association members conduct and support research about the Lincoln Highway and publish a magazine, "The Lincoln Highway Forum," for articles on the history of the highway and news about the association. They also work with local communities and businesses along the Lincoln Highway route to promote the highway as a tourism destination.

For more information about the Lincoln Highway Association, or to become a member, write to: Lincoln Highway Association, P.O. Box 308, Franklin Grove, IL 61031 or visit the Association's website at: www.lincolnhighwayassoc.org

There are active state chapters in New York, New Jersey, Pennsylvania, Ohio, Indiana, Illinois, Iowa, Nebraska, Wyoming, Utah, Nevada and California.

Author Larry Krug stands by concrete Lincoln Highway marker in front of the Lincoln Highway National Headquarters in Franklin Grove, Illinois

Lincoln Highway Marker

Wednesday, September 17
Day 9

Left Trenton, New Jersey at 9:30 a.m.

Log: Stops at Philadelphia, Coatsville, Lancaster and York, Pennsylvania. 95 miles. Spent night at Colonial Hotel in York.

Review: Big luncheon given to us at Hotel Bellvue in Stratford by Women's Coolidge Club.

Large line of cars escorted us all day long.

Special banquet given in our honor by Republicans of York.

15 mile parade of cars from Lancaster into York where 55,000 people jammed the streets to see the caravan. This is the largest crowd of the entire trip.

C. E. Thompson left us in Philadelphia, and we were joined by a new Lincoln driver.

J. M. McCullough of the Philadelphia "Inquirer" now with the party.

Parade and meeting in York lasts until midnight.

—E. A. Seidel

Small towns along the way were equally enthusiastic about the traveling Republican caravan. The following article appears in the September 18 edition of the *Coatesville Record*:

Coolidge-Dawes Caravan Given Rousing Greeting On Trip Through County; Loyal Republicans Wait Hour in Rain To Meet Party At Stafford; Former Chester Countian, Who Is Directing Tour, Meets Relatives and Old Friends At Downingtown And Coatesville.

Despite a chilly rain and wind that made the day miserable, a rousing welcome was extended by Chester County people yesterday to the Coolidge-Dawes Lincoln Highway caravan which enter the county at Strafford shortly before three o'clock and left it near Black Horse a little more than an hour later after short meetings had been held at Downingtown and Coatesville.

The passing of the caravan was of more than the usual interest to Coatesville for the reason that it marked the home coming of a Chester County boy who left his home near this city thirty years ago to carve out his fortune in the west. So successful has he been that his home city in Omaha sent him to the national Congress a few years ago and he is now taking a leading part in the Republican National Campaign. This man is Albert W. Jefferis, director of the tour.

The feature of the passing of the caravan was therefore the speeches of Mr. Jefferis in Downingtown and Coatesville and especially that made in Coatesville where he referred to his boyhood here and to the love that he still cherished for his native county. Here also he met his three brothers, Barton E. Jefferis, H. Walter Jefferis and Carl S. Jefferis, all of whom met the caravan at Strafford. In Downingtown he met a sister, Mrs. Mary R. Swing, who teaches in the High school there. Another sister, Miss Esther A. Jefferis, of Romansville was unable to meet him. Others present were Mrs. Gertrude E. Jefferis, Miss Elizabeth Jefferis, Mrs. H. Walter Jefferis and Mrs. Carl S. Jefferis.

The caravan which was to have entered the county at Strafford at 1:50 was more than an hour late, but the twenty-five automobile loads of enthusiastic Republicans who went there to greet it waited patiently in the rain until it came. Virtually every section of the county was represented in the group. From Coatesville there were Mayor A. H. Swing, Representative William W. Long and State committeeman W. E. Greenwood. From West Chester there were State Senator T. Larry Eyre, County Chairman William H. Clark, Vice Chairman Mrs. E. Page Allison and many others. Still others included Mrs. Charles Wister Ruschenberger, of Strafford, "Squire" Eaves, of Paoli, George B. North, of Lyndell, and Rev. Thomas R. McDowell, of Elkview.

The caravan consisted of six automobiles one of which was a truck on the sides of which was painted a picture of President Coolidge's birthplace at Plymouth, Vt., with an electrically illuminated sign on the front containing the slogan, "Common Sense and Brass Tacks." The truck also carried the pictures of President Coolidge and General Dawes. Among those in the caravan party was Albert W. Jefferis, Jr., a son of Congressman Jefferis, and Herbert L. Moore, of Plymouth, Vt, a boyhood friend of President Coolidge and a typical New England farmer.

After the caravan arrived at the county line no time was lost in lining up the procession. Each automobile in the delegation was supplied with a Coolidge-Dawes tour pennant and with Senator Eyre leading the way and Representative Long bring up the rear the procession started up the pike. All along the way to Downingtown people were out along the highway to see the procession go by and to wave a greeting. At every school house the teachers had the children lined up in front of the buildings and in some places they waved American flags to the passing tourists.

The first stop was made in Downingtown where a large crowd had gathered in front of the Swan Hotel. Senator Eyre escorted Congressman Jefferis to the hotel porch and introduced him to the crowd. A cheer went up as the crowd got its first view of the former Chester Countian. Congressman Jefferis is a large well built man with a great shock of almost white hair, a full determined countenance and a big voice. Attired in a large black beaver hat, flowing black tie, gray trousers and spats and a long coat he looked every inch a barnstorming westerner and he captured the crowd at once.

Following a short speech in which he emphasized the fact that President Coolidge's election in November is of vital importance to every voter for the reason that he is pledged to economy and lower taxes, the caravan started for Coatesville with Representative Long in the lead. Mayor Swing had telephoned from Strafford to have the street cleaned from Third avenue to the post office building in order that the procession might have places to park their cars when they arrived. The plan worked out like clock work and when the caravan arrived the cars simply slipped into the stalls along the north side of Lincoln Highway and there was little or no confusion.

The same crowds that had greeted the caravan from Strafford to Downingtown appeared along the route from Downingtown to Coatesville and in this city several thousand people crowded the sidewalks as the procession moved along. In the presence of a crowd that completely blocked the street, Representative Long introduced congressman Jefferis, from the post office steps, as the man who had nominated General Dawes for Vice President in the convention and later notified him of his nomination. The former Chester Countian faced the crowd amid a burst of handclapping that quickly subsided as he began to speak.

After a few preliminary remarks in which he told of his boyhood days in Chester county and the fact that he had left here when he was twenty-three years of age, Congressman Jefferis launched into his address. He told how taxation affects every citizen of the Republic whether he pays any direct taxes or not and how the policy of President Coolidge in seeking to effect economy and reduce taxes will benefit every citizen. He made a plea that his native county stand firm in her Republicanism in November and ended by wishing Coatesville and this end of Chester County the greatest measure of prosperity they had every enjoyed.

Congressman Jefferies left the post office steps immediately to mingle with the crowd and shake hands with his brothers and other old friends. This required but a minute or two and then the caravan was in motion again. The Chester County delegation escorted it to Sadsburyville where it was met by Congressman Griest, Senator Homsher and other prominent Republicans of Lancaster. After a stop of about five minutes in the village, the caravan moved on to the Lancaster County line just beyond Black Horse where it was met by a delegation of several hundred Lancaster County Republicans who escorted it through that county.

When the caravan reached Sadsburyville a supper given by the Sadsburyville Fire Company was in progress at the Fire House and a number of the visitors made a raid on the supper tickets. Few of them remained, however, to eat but returned to Coatesville almost immediately.

The local committee for York and York county did an exceptional job of preparation for the caravan. With police cooperation all streets along the parade route were cleared of parked vehicles. Cary Etnier, chairman of the general committee, called for an open invitation to all citizens of York county to participate in the parade of automobiles by decorating their vehicles and joining the parade from side streets as it passed by. The Ford Motor Company contributed the use of three large automobiles which were used to carry the escort at the head of the parade.

This attractive Coolidge trigate button and ribbon attachment was distributed by the Fremont Lincoln Coolidge Veteran Club of Lancaster County, Pennsylvania. 2-1/4" button, 7" overall length

Large 10-1/2" Caravan ribbon with button attachment

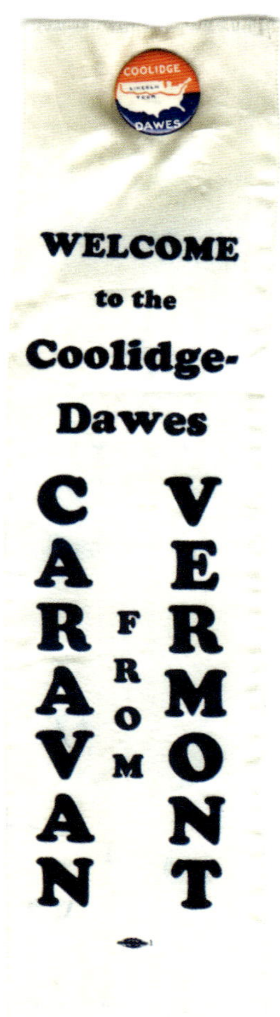

Herb Moore holding an audience in the rain during a Philadelphia stop (newsprint)

Colonial Hotel, York, Pennsylvania

The York Dispatch fully covered plans for the event in two columns the day of the caravan visit, including the complete parade route: East Market street to Penn street, Penn street to Roosevelt avenue, Roosevelt avenue to Madison avenue, Madison avenue to Richland avenue, Richland avenue to King street, King street to South George street, and South George street to Center square.

Lengthy coverage of the caravan's visit was reported in *The Gazette* and *Daily Article*, in York: **"Immense Crowd Attracted by Novel Republican Caravan; Coolidge-Dawes-Lincoln 'Overland' Parade Is Late In Arriving; Gets Fine Reception From York Citizens."**

The Coolidge-Dawes-Lincoln caravan was enthusiastically received last night by throngs of York citizens on its arrival here. Its late arrival did not prevent thousands of persons assembled in the downtown section, and on streets over which the caravan followed by many automobiles filled with York countians passed, waiting until the last car in line had gone by.

The birthplace of President Coolidge as pictured on a large truck in the line caused some attention, but the decorated motor cab occupied by a number of York citizens who voted for Abraham Lincoln, was the objective of all eyes.

The parade got started late because the caravan was held up by the Lancaster county reception and never arrived at the Valley Inn, until after 7:30 o'clock. Here a fine dinner was served to the members of the party and the Reception committee. The committee met the caravan at Columbia. It was 8:25 o'clock before the party left the Valley Inn.

Chief Marshal John W. Richley had his plans well laid, however, and as the automobiles began to move and string out, others gathered in numbers on lanes and streets bordering on the Lincoln Highway also fell into line. It was 9 o'clock as the procession passed through center Square and after a run over the route of parade as previously unannounced, speechmaking was in order in the square. There were but two speakers, Former Congressman Jefferis and Herbie Moore, a boyhood friend of Cal Coolidge. They were presented by Attorney R. E. Cochran.

The streets were jammed with a mass of humanity as the parade slowly moved onward. Many residences were ablaze with light. Members of the party were confident it was the finest reception which they have received since starting on the coast-to-coast boost for Coolidge.

As the caravan and its escort, enroute from Columbia to the Valley Inn, approached that historic spot a weird site greeted the members of the party. In a field just east of the Inn, and north of the highway, stood seven figures in the full regalia of the Ku Klux Klan. They stood motionless, presenting the appearance of a guard of honor, the central figure holding aloft a beautiful American flag.

To the disappointment of the great crowd gathered in Center Square to hear the speechmaking the amplifying device which it was advertised the party carried with it, thus enabling persons on the remote fringe of a great crowd to hear the speakers, was not in evidence. Members of the party stated that it had broken down. They didn't have very good luck with it since starting on the tour. As it was difficult to hear the speakers last night, the crowd soon thinned out.

The local committee of arrangements was on the job and as early as 4:30 o'clock left York to go to Columbia to meet the visitors, and escort them here. The caravanists themselves were responsible for the long delay, having become the central figures in a great demonstration in Lancaster city and county.

This morning the caravan leaves York for Hanover, the home of Ward C. Haffner, the Coolidge-Dawes elector in this district. The schedule calls for arrival in Hanover at 9:30 o'clock. Other places to be reached are – Abbottstown, 10 a.m.; Gettysburg, 10:30 a.m.; Chambersburg 12 o'clock noon.

Hon. A. W. Jefferis, who is a former member of Congress from Nebraska, placed Gen. Charles G. Dawes in nomination for the vice presidency at the Cleveland convention. He spoke last evening of the simplicity of the life of Calvin Coolidge, drawing many parallels between the President and Abraham Lincoln. The rise of Coolidge, he declared, was an exemplification of the opportunities offered in America. In the hands of such a man, standing on the principles of the Republican platform, he said, the country could be assured another four years of prosperity and sane government.

Referring to the issues before the people at the coming election Mr. Jefferis said: A lot of people have allowed their feet to leave the ground in this campaign. Their heads are filled with a lot of strange fantasies and they have been guided in this direction by the misguided minds of a lot of politicians and demagogues. But just look around you at these teeming fields in old York county. Are the people idle? Are you discontented? Are you ready to abandon the sound doctrines of Republicanism for the so-called progressivism of Robert M. LaFollette? I don't think so. Do you think that the American people want to try another experiment of Bryanism by voting the Democratic ticket which has such a conspicuous Bryanite appendage?

This is the anniversary of the adoption of the constitution of the United States, and so it is the anniversary of the greatest day in the progress of the ideals of human liberty. It is fitting that in this historic town where the Continental Congress sat in the darkest days of our fight for American independence that we should recall the making of that great instrument for civil liberty and urge our fellow citizens in every state to stand up for the support of the Constitution under the administration of two such men as Calvin Coolidge and Gen. Charles G. Dawes.

Herbert L. Moore, a native of Plymouth, Vt., was the boyhood playmate of Calvin Coolidge. They lived on adjoining farms. Calvin went to Amherst College and became a lawyer. Herb Moore went to the academy at Ludlow, Vt., and returned to the farm. He is still a farmer, and is perhaps the only "dirt" farmer in this campaign who has gone on the national stump. Without any of the grace of oratory, he tells with the eloquence of sincerely and straight-forward expression, the story of the boy Coolidge. Everywhere he has won his audiences with his quaint stories of life in the mountain of his native state and how from those simple surroundings he rose to the office of the world's greatest executive.

To his audience in York Mr. Moore's best hit was his story of Cal Coolidge's ability as a milker of cows and in his judgment in picking out the most productive sugar maples on the Vermont farm.

George Washington Shock, 247 South Pine Street, one of the men who rode in the motor cab last night occupied by men who voted for Abraham Lincoln also served his birthday anniversary yesterday. Mr. Shock voted for Lincoln in 1860 and 1864. He was privileged to shake hands with President Lincoln. Mr. Shock was born on September 17, 1838.

Other Lincoln voters in the parade last night are: Thomas A. Benton, 46 East Princess street; Lewis Schriver, 761 West King street; Hugh W. McCall, 316 East Market street; Frank Ginter, 29 South Newberry street; A. R. Feistel, 441 Lincoln street, and Jonathan Jessop, 313 East Market street.

Lincoln voters who did not take part in the procession but who have asked for permission to sign the register are John J. Frick, 101 Springettabury avenue; Matthew H. McCall, 488 Linden avenue; Robert Arthur, 302 North Hartley street; William A. Lowe, 37 West Maple street; Charles F. Ford, 218 West College avenue; Emanuel G. Martin, 756 West King street.

Thursday, September 18
Day 10

Left York, Pennsylvania at 9:00 a.m.

Log: Stops at Gettysburg, Chambersburg, McConnellsburg, Port Souden, Everett and Bedford, Pennsylvania. 105 miles. Spent night at Hotel Pennsylvania in Bedford.

Review: Crossed the Tuscarora Peak, elevation 2,905 feet.
Visited the Battlefield of Gettysburg and signed up the old veterans who voted for Abraham Lincoln. This book, when finished, will be presented to President Coolidge.

The oldest one signed up thus far is E. J. McElwain of Holyoke, Massachusetts, aged 96 years, who not only cast his ballot for every Republican presidential candidate but has the distinction of having voted for General Winfield S. Scott, the last Whig candidate in 1852.

—E. A. Seidel

In both the preliminary coverage of the caravan's visit to Chambersburg, and the actual coverage of the event, the *Chambersburg Public Opinion* made much mention of the Lincoln registry: **"Local Men Who Voted for Lincoln Are to Register In Book Carried by Coolidge Tourists Here Tomorrow, Meeting in Square at 12:45 Noon"** was the heading of September 17 article, and **"Seven Lincoln Voters Sign As Caravan Visits"** was the heading of September 18 paper.

Enthusiasm which has marked the reception accorded the Coolidge-Dawes caravan on its cross-continent tour was sustained here today in a large meeting in Memorial Square, presided over by Charles Walter of town and addressed by Congressman A. W. Jefferies of Omaha, Neb., and Herbert L. Moore of Plymouth, Vt.

Forty automobiles formed the procession, which was met at East Point by the Queen City Band and escorted to the band stand in Memorial Square. The automobiles, gaily bedecked in pennants, banners, flags and Coolidge stickers presented a patch-quilt effect as they coursed through the crowds which had gathered along Lincoln Way East and in Memorial Square.

Those composing the party were Congressman Jefferies, director of the tour, John P. Cowan of Pittsburgh, former newspaper man and publicity director for the Pittsburgh Chamber of Commerce, Herbert L. Moore of Plymouth, Vt, A. W. Jefferies, Jr., William Wallace of Omaha, Neb., S. E. Moliden, E. A. Seidel, E. C. Hoch, all of New York City, Frank Stearns of Philadelphia, and two newspaper men, T. McCullough of the Philadelphia Inquirer and L. Lyons of the Boston Globe.

Addresses were made in Memorial Square by Congressman Jefferies and Mr. Moore. In his talk, which was mainly of a patriotic nature, the Nebraska congressman stressed the opportunities which the country offers to the youth of the nation and urged a continuance of Republican policies under which he said the country has prospered. Mr. Moore distinguished between the acceptances addresses of President Coolidge, John W. Davis, the Democratic nominee for president, and Senator LaFollette, Progressive nominee, as a means of showing the campaign attitude and policies of the various nominees.

Following the meeting the Lincoln register was signed by seven local men who voted for Abraham Lincoln.

Those who registered were Dr. George Fiske Platt, who voted for Lincoln in 1864, while a resident of Chambersburg.

H. P. Bingham of 133 Norland avenue, who voted for Lincoln twice, in 1860 and '64. Mr. Bingham was a member of the body guard for President Lincoln at Gettysburg the night before he delivered his famous address. At that time, Mr. Bingham was a member of the 21st Pennsylvania cavalry.

Capt. S. I. McPherran of 580 Nelson street, who voted for Lincoln while a member of the army in 1864, going to his home in Mercer, PA, to vote.

B.F. Zook of 117 N. Main street, who voted twice for the martyred president.

D.K. Appenzellar of Lincoln Way East, Joseph Sollenberger of North Main street, Robert Dunkinson of Harrison avenue. Mr. Dunkinson cast his ballot for Lincoln while on the railroad train as a member of a Pennsylvania artillery unit bound for Pittsburgh.

John Harvey and Martin Forney of town also voted for Lincoln but their signatures were not secured.

Hotel room at Hotel Pennsylvania in Bedford, Pennsylvania. Driver Seidel drew some "handiwork" on the picture

Friday, September 19
Day 11

Left Bedford, Pennsylvania at 9:00 a.m.

Log: Stops at Jennerstown, Ligonier, Greensburg and Pittsburgh, Pennsylvania. 96 miles. Spent night at Hotel Henry, Pittsburgh.

Review: Rained all day yet large crowds greeted us everywhere. Met outside of Pittsburgh by 300 machines and a band of music.

Have decided to lay over in Pittsburgh until Monday morning, as KKK are holding a meeting in Canton, our next night stop, and we do not wish to conflict with them.

Meeting at Jennerstown held in pouring rain, yet the 2,000 people who had gathered stood for 45 minutes and listened.

—E. A. Seidel

The *Greensburg Daily Tribune* carried a full front page headline **"Throng Greets Republican Caravan Today"** as a special report from Ligonier.

The group of Republican county men and women on their way to the Somerset-Westmoreland county line, met the Coolidge-Dawes Caravan on schedule at 12 o'clock. The party is now on its way to Greensburg, having stopped off here (Ligonier) for a few minutes.

An enthusiastic greeting was given the tourists. There were about 40 automobiles loaded with about 200 people on hand for the welcoming. Many hundreds more saw the procession in Ligonier.

The Republican campaign for national, state and local candidates gets under way in Greensburg today. Headed by the group of men on the Coolidge-Dawes Lincoln tour, a delegation of Pittsburgh and Allegheny county politicians and hundreds of prominent Republican men and women from Westmoreland county gathered here this morning. A short meeting at the Penn Albert was the schedule.

Preceding this afternoon's meeting, about 75 automobiles filled with men and women left Greensburg this morning for the county line. There the Coolidge-Dawes touring party was met and the trip to Greensburg started. The procession attracted much of attention as it wended its way over the hills of the Lincoln highway toward this city. A short stop was made in Ligonier. There, many veterans of the Civil War and other aged residents who cast their first vote back in 1861 for Abraham Lincoln were given the opportunity to sign the Lincoln register. This book is later to be presented to President Coolidge. All along the route there were gathered in small groups people who had read of the passing of the caravan.

Original arrangements called for the welcoming party to leave the court house at 9:30 o'clock this morning. In a late hour conversation over the telephone last night, J. Arthur Thomas, Republican county chairman, learned that the tour had been delayed on its trip. This made the hour of the arrival in Greensburg more or less uncertain. It would be an hour later, at least, it was stated.

Early this afternoon many people assembled at the Penn Albert hotel to get a glimpse of the tour. Pittsburgh and Allegheny county men left their city about 11 o'clock this morning to greet the caravan here and stay for the local meeting. They were then to escort it to Pittsburgh where elaborate arrangement have been made for a demonstration late this afternoon. A large mass meeting will be held in that city tonight.

Prominent Republicans were among those who made the trip this morning to the county line. Every one of the county candidates made the trip. Cyrus E. Woods, ex-ambassador to Japan was also among the group. Senator Woods is the presiding officer at this afternoon's meeting. Chief Burgess Harry N. Yont is to give the address of welcome to the tourists.

Early this morning groups of Republicans from practically every town and borough in Westmoreland county were on their way to this city to join the welcoming party. Others rolled into position along the line. They were ready to join the caravan when it reached the convenient place to them. Monessen, New Kensington, West Newton, Jeannette, Lathrobe, Scottdale and other places all contributed their quota. There was at least one automobile load of women from each of the Greensburg wards. There were many other women in the line.

Grand Army veterans turned out in a good assemblage this morning. Automobiles were provided to take them to the county line. Whether it rain or not, made little difference to these grizzled warriors.

Among those who made the trip this morning were: Quigley Waters, J. E. Stephenson and Postmaster John McCune of West Newton; Maude Trescher, Mrs. W. D. Hunter, Mrs. James Baugher, C. M. Bomberger, Jeannette; C. C. Walthour, Bennett Kiehl, John E. Kunkle, A. I. Hurning, John V. Stephenson, Paul W. Abraham, Miss Elizabeth Sweeny, Miss Laura Cope, Congressman Adam M. Wyant, G. A. Smith, Derry; Rev. J. F. Stray, Lathrobe; Howard F. Reider of New Kensington; Oliver M. Deibler, Paul K. McCormick, Miss Carolyn Jamison, County Chairman J. Arthur Thomas, Harold A. Stewart and Senator James B. Weaver, of Lathrobe; Attorney Clifford M. Smith of New Kensington, M. P. Endean and John Endean, of the same place; J. R. Smith, B. T. Kiehl, J. E. Tracy, James Keating, E. A. Cremer, Elliott L. Hibbs, James F. Torrence, H. F. Bovard, C. C. Foight, W. C. Henderson.

The Civil War veterans who made the trip were E. B. Sweeny, Nicholas Long and Frank Coleman, of New Stanton, Joseph Shrader, W. F. Weaver, J. W. King, W. T. Maxwell, J. H. McCelland, Frank Colvin, George Weddell, Paul Henry and Silas Frye.

This Coolidge-Dawes ribbon, with a Lincoln medallion hanger, was distributed by the Lincoln Club of New Brighton, Pennsylvania. 6-1/2" long

The next morning's *Greensburg Morning Review* (September 20) covered the speeches made during the caravan's visit to Greensburg in a front page feature with the heading: "**Coolidge-Dawes Trip is Given Big Reception in Old Westmoreland.**"

"Keep cool and keep Coolidge," was the parting admonition of H. L. Moore of Plymouth, Vermont, neighbor and friend and classmate of President Calvin Coolidge, Republican nominee for the presidency of the United States, as he closed an address made to a large crowd on the Penn Albert Plaza yesterday afternoon.

Mr. Moore was the first of the distinguished visitors of the Coolidge-Dawes Lincoln tour to address the Greensburg audience as they remained over in the Westmoreland county seat for a short time, on their journey across Pennsylvania toward the Western coast. He was introduced to the gathering by Senator Cyrus E. Woods, Greensburg, distinguished fellow townsman, and former Ambassador to Japan, who presided during the meeting. Mr. Woods was introduced to his neighbors by J. Arthur Thomas, Republican county chairman, and the ex-ambassador made a cordial address of welcome to the travelers. A word of welcome was also voiced by Burgess Harry N. Yont.

The Coolidge-Dawes-Lincoln tour rolled into town slightly behind their schedule and the members of the party immediately disembarked for luncheon at the Penn Albert. The crowd gathered early and in a short time, completely filled the parking place and the plaza on the north side of the hotel. Seventy-five machines from the county seat and neighboring towns had carried interested Republicans to the eastern extremity of the county to form an escort of honor into Greensburg, and these also disgorged their passengers. The crowd was large and interested. Women formed a fair proportion of the audience...

Former Congressman A. W. Jefferis was the next and concluding speaker...the voice of Mr. Jefferis was uncertain at the outset and by way of apology for his croaks and quavers – which later disappeared entirely – the Nebraskan told of the ten days already spent on the trip. The tour opened in Plymouth, Vermont, and will continue across 17 states to the Pacific coast. From 10 to 15 hours have been consumed a day, and the purpose, so Mr. Jefferis said, is to consider one or two men that a town of 30 people can produce.

Plymouth, he said, has produced a president of the United States. It has few telephones, no electric lights and is 12 miles to a railroad. I think I know why it is. There one finds devotion and loyalty to each other. When Mr. Moore here heard of the untruths being told about his boyhood playmate he left his farm to trail through the country on this tour.

Mr. Jefferies turned his verbal guns on Robert LaFollette of Wisconsin, on the group of radicals he leads, and on the party he has created, predicting that if elected, LaFollette never could carry out his promises because of the widely differing principles of the groups he has attempted to weld into a party. An engine whistled loudly and long on the tracks of the Pennsylvania railroad nearby. Seizing upon the significance of the noise, the speaker quickly dealt out figures showing the country's wealth and prosperity. He said: "The United States produces 46 percent of the world's output of coal and uses 42 percent; produces 54 percent of the iron and uses 53 percent; produces 64 percent of steel and uses 47 percent; produces 49 percent of the copper and uses 44 percent; produces 60 percent of the cotton and uses 37 percent; produces 52 percent of the shoes and wears 51 percent on the feet of her people. Oh, there is something doing in old America" he concluded, as the whistle gave a final shriek and the train passed on.

After roundly scoring the doctrines of discontent preached by the radicals, and making a plea for the upholding of the constitution and all the present triumphs enjoyed under it, the speaker paid a fine compliment to Congressman Adam M. Wyant of Greensburg. He said: "You have a member of Congress here with whom I served four years. I want to tell you that Adam Wyant stands are high with that body for they know he is a clear thinker. He like all people is interested in seeing happiness prevail in the country. Don't forget him. Do your duty there also."

Attractive window sticker promoting Pennsylvania candidates along with Coolidge and Dawes. 5" x 7-1/2"

THE CALVIN COOLIDGE MEMORIAL FOUNDATION

The Calvin Coolidge Memorial Foundation, Inc. (CCMF) was established in 1960 in part to redress the fact that there is no presidential library for Calvin Coolidge, our 30th U.S. President. The Foundation develops and publishes materials and programs to preserve Coolidge's legacy, and provide primary and interpretive information through their website and by request, to the media, students, scholars, genealogists and the general public. The Coolidge Foundation owns and maintains its offices in the 1840 Union Christian Church in Plymouth Notch, Vermont.

With the exception of the Union Christian Church, all the other buildings at the Historic Site are owned and operated by the State of Vermont Division for Historic Preservation. Visitors come to Plymouth Notch to visit this unspoiled site where Calvin Coolidge was born and raised, and where he was sworn in as President by his father John Coolidge, on August 3, 1923.

For more information about the Calvin Coolidge Memorial Foundation, or to become a member, write to: Calvin Coolidge Memorial Foundation, P.O. Box 97, Plymouth, VT 05056 or visit the Foundation's website at: www.calvin-coolidge.org

Union Christian Church, Plymouth Vermont. Erected in 1840

President Coolidge at Plymouth, Vt.

Saturday, September 20
Day 12

Lay over in Pittsburgh, Pennsylvania

"Due to the condition of the automobiles carrying the party and the necessity to stop for overhauling the machines, and also to the demands being made upon speakers with the touring campaigners, the Coolidge-Dawes transcontinental caravan which is proceeding over the Lincoln and Harding highways will remain in Pittsburgh over Sunday," was reported in papers along the route.

Sunday, September 21
Day 13

Lay over in Pittsburgh, Pennsylvania

Official Caravan pennant. 30" long

Monday, September 22
Day 14

Left Pittsburgh, Pennsylvania at 8:30 a.m.

Log: Stops at Beaver Falls, Pennsylvania and Columbiana, Salem, Alliance and Canton, Ohio. 100 miles. Spent night in Canton.

Review: Syd Lewars, Lincoln driver, left us at Salem and we were joined by G. A. Rickenbaugh of the Cleveland branch.

Senator Fess rejoined us at Pittsburgh to campaign through Ohio.

Presented with Ohio State flag at Columbiana.

Motorcycle escort overturns outside Salem while doing 50 miles an hour. Side-car fell on him but he was not hurt.

—E. A. Seidel

After a morning meeting in Beaver Falls the caravan moved on toward the Ohio state line. Near East Palestine they were met by a large caravan of autos headed by the Columbiana county central committee, of which County Auditor C. E. Hamilton of Lisbon is chairman. The Ohio flag was presented by Assistant Prosecuting Attorney R. M. Brookes of East Liverpool. The entourage then proceeded on their way.

In Columbiana, where the caravan arrived about 1:30, the tourists had their lunch, and then began the short trip to Salem.

The September 22 edition of *The Salem News* reported that the caravan arrived in Salem shortly before 3 p.m., making their first scheduled Ohio stop.

At the city limits city officials and police met the caravan and the escorting automobiles. The Quaker City band also was present. Because of the uncertainty of the arrival of the Lincoln caravan, there was not a large crowd out to welcome the tourists. Atty. H. L. McCarthy, chairman of the speakers committee, will preside and Probate Judge Lodge Riddle and Atty. W. H. Vodrey of East Liverpool, other members of the committee, will speak. The program was to be followed out as previously arranged, and at press time, speakers were preparing to give short talks to those who congregated to greet the caravan. After the program here, the caravan was escorted to the city limits at the Damascus Road and turned over to the Canton committee, which with the sheriff, will escort the party into Alliance and Canton.

Senator Simeon D. Fess, acting as official host of the state of Ohio, will travel with the group and give short talks at the stopping places.

The following newspaper coverage of the local plans for the caravan as covered in the *Alliance Review* shows the operations of a local stop and the planning that went into it. First reported on September 13, **"City On Route Of Coolidge-Dawes Lincoln Tourists."**

Alliance is on the itinerary of the Coolidge-Dawes "Lincoln Tourists" now headed westward on a coast-to-coast stumping trip.

Included in the party of political speech-makers will be A. W. Jeffries of Omaha, Neb., the man who delivered the notification address to Brigadier General Charles G. Dawes, vice presidental nominee.

H. Ross Ake of Canton, chairman of the Stark County Republican Central committee, today issued a call for an executive committee meeting at Canton Monday night to complete plans for welcoming the party in Stark county.

The party, consisting of several automobile loads of dignitaries, is scheduled to leave Pittsburgh Saturday morning, September 20. It will make a brief stop at Beaver Falls and arrive in Salem for noon.

Proceeding westward over the Lincoln Highway, the party will stop here for a brief meeting before proceeding on to Canton.

The September 16 issue carried a story, **"G.O.P. Caravan To Get Big Greeting In Stark County."** The article read:

The "Covered Wagon" will be modernized this week in Alliance when a caravan of pilgrims from the quaint Vermont village of Plymouth pass through the city in an ocean to ocean journey to extol the name of their favorite son, Calvin Coolidge, President of the United States and Republican candidate for re-election.

The Coolidge-Dawes Lincoln tour is the official name of the caravan and the purr of powerful motors will replace the hoof beats of the steeds that once drew caravans from coast to coast. It is slated to reach Alliance Saturday afternoon around three o'clock en route from Salem, where the party will stop for lunch, to Canton, where it rests at six o'clock. But two stops will be made in Stark County.

The event will be the opening of the Republican presidential campaign in Stark county. Preparations to receive the caravan in fitting style were made at the meeting of the Stark county Republican executive committee at a meeting held last night.

Between 50 and 100 automobiles filled with Canton and Alliance people will welcome the caravan and escort it across Stark county. Senator Fess will join the party as it crosses the Ohio line and make the trip across Ohio. The caravan will follow the Lincoln highway across the United States to the Pacific coast.

The caravan, which consists of six machines, is under the direction of A. W. Jeffries, former Nebraska congressman. John B. Cowan is manager of the tour. A national speaker will make the trip with the caravan.

The remainder of the touring party is composed of residents of Plymouth who belie the adage that a prophet is not without honor save in his own country.

Two special features of the caravan make it unique. A reproduction of the Coolidge home will be shown on one of the automobiles. A book will be carried in which all men who voted for Abraham Lincoln will be asked to sign their names. This book will be given to the President at the conclusion of the tour.

The following day, September 17, it carried the article, "**Alliance Will Send Party To Meet Stumpers**" which read, in part:

A cavalcade of automobiles will drive away from Alliance shortly after noon Saturday to meet the Coolidge-Dawes Lincoln tourists on the border of Stark and Columbiana counties, if present plans are carried out.

Alliance spokesmen on the Stark county G.O.P. committee – Miss Evelyn Vale, Sydney Geiger and W. A. Thompson – have taken the initiative in drafting tentative plans for welcoming the coast-to-coast tourists. It is proposed that an open meeting be held, likely on the square, during the brief stay of the Coolidge-Dawes party in Alliance..."

The September 20 front page of the *Alliance Review* carried the headline, "**G.O.P. Plans Welcome To Stumpers On Visit Here.**"

Reorganized for the coming campaign, the Alliance Republican central committee at a meeting last night completed plans for welcoming the Coolidge-Dawes Lincoln tourists on their coast-to-coast trip here Monday afternoon.

George R. Floyd was re-elected chairman and Attorney F. E. Hunter, secretary, of the Alliance G.O.P. committee at the Reorganization meeting held in Attorney S. L. Geiger's office. Plans for an active campaign were laid.

Opening gun in the fall campaign here will be fired Monday afternoon when a mass meeting is staged on Public Square as a welcome to the Republican stumping party. A cavalcade of automobiles will form on the square at 1 o'clock Monday afternoon and drive to Salem to meet the Coolidge-Dawes party.

It is likely that the Reed Boys' band will be taken along to furnish music for the welcoming party. The meeting on the square will begin at 3:30 with George R. Floyd as chairman. Senator Simeon D. Fess, who is escorting the tourists across the state, will be on the speaking program. In case of inclement weather the meeting will be held in the council chamber at the City Hall.

Campaign plans tentatively laid by the Republican Central committee last night call for at least two big mass meetings here before the fall campaign curtain is rung down. A state meeting and a national rally are planned."

And, finally, the September 23 issue, the day after the Alliance visit, reported, "**Alliance Greets Coolidge Caravan On Tour Across Continent; Hear Able Speeches As Campaign Opens.**"

Alliance greeted the Coolidge-Dawes coast to coast caravan late Monday afternoon, as the party arrived from Salem escorted by autos from Alliance and Canton.

Senator Simeon D. Fess, ex-Congressman A. W. Jeffries and T. C. Hunsicker were the speakers.

I will never vote for an alliance with any European country until it bows to America in submission, declared Mr. Hunsicker in a stirring appeal to the voters not to destroy the great privilege they have been enjoying. He urged that President Calvin Coolidge be elected in the interests of preserving national prosperity.

A dozen or more automobiles filled with Alliance people met the caravan at Salem and escorted it to the public square where a large delegation from Canton joined the party.

George Floyd as chairman of the meeting introduced each of the speakers in turn. Mr. Floyd is chairman of the Republican central committee.

The caravan arrived in Alliance enroute from Salem to Canton more than an hour late. The meeting on public square did not break up until nearly 6 o'clock.

Senator Fess once more sounded a warning he has previously voiced that peculiar conditions in the present campaign may result in the election to the presidency of Charles W. Bryan.

Don't be deluded into the belief that the next president must be either Coolidge or Davis, he asserted. "If the next president is not Coolidge it is likely to be neither Davis nor LaFollette but Bryan."

With the close of the World War it was the enactment of a protective tariff that saved the country, Senator Fess declared. He told of the accomplishments of the Harding administration in reducing the public debt.

"There is no need for America to join in a conference called by the League of Nations," the senator exclaimed. "We have reduced our army from 4,500,000 men to 127,000 men. Thus we have met the example without a conference and if there is to be any meeting in which America is to have a part it must be held here, away from the rivalries and jealousies of Europe."

He lauded the work of Secretary of the Treasury Mellon, asserting Republican administration had reduced the public debt from 24 billion to 21 billion dollars, cut the taxes from five billion to less than four billion dollars in a year, a saving of $4,000,000 daily and reduced the interest rate from 5 3/4 percent to 4 1/2 percent, a saving of $40,000,000 a year in interest.

Senator Fess characterized Secretary Mellon as having shown more brilliance than any man who has held the office since Alexander Hamilton.

He touched upon America's foreign relations asserting that today America's foreign relations are on a higher plane than ever before. "All our disputes have been adjusted and we are on terms with everybody but Russia and we will have nothing to do with the Russians until they are willing to meet us on American standards of honesty." he declared.

An evening meeting was held in Canton. The header used by the AP story, "Fess Raps LaFollette," kept up the barrage of attacks on LaFollette by Republican speakers with the caravan.

Senator Robert M. LaFollette, independent candidate for president, was the object of attacks by Senator Simeon D. Fess and A. W. Jefferies of Omaha, Neb., members of the Coolidge-Dawes trans-continental tour party, at a meeting here (Canton) last night following the arrival of the caravan from Alliance. both speakers emphatically suggested that all persons not satisfied with America and its form of government should take up their residence in Russia.

Senator Fess made a plea for a continuance of the protective tariff system. He lauded the assertion of President Coolidge declaring that the handling of the financial problem has been the most brilliant in the history of the country and the secretary of the treasury, Mr. Mellon, has exhibited financial genious second only to Alexander Hamilton.

Tuesday, September 23
Day 15

Left Canton, Ohio at 9:15 a.m.

Log: Stops at Wooster, Ashland, Mansfield and Bucyrus, Ohio. 108 miles. Spent night at Hotel Elberson, Bucyrus.

Review: Dinner at Elk's Club.
—E. A. Seidel

The September 23 issue of *The Mansfield News* reported, "**Coolidge Party Is Here Today; Meeting Ready**," giving details of local Republican plans including the caravan parade and open-air meeting scheduled for Central park.

At the meeting Congressman Jefferies served as chairman and spoke of the wonderful reception the caravan has been getting at every point along the line. Herb Moore extolled the virtues of Coolidge as a common man. Senator Fess of Ohio again made his bitting attack on LaFollette. Bitterly the Ohio senator attacked the onslaught which Sen. LaFollette and his supporters have made against the president and the administration. "There are men in America today who are deliberately trying to tear down the Constitution of your country and mine," he declared. "We will have no Russia here but I wish to God that these men would go to Russia where they belong."

With such a ticket as this headed by Calvin Coolidge and Charles G. Dawes the Republican party cannot do otherwise than win this fall," declared the Ohio senator. "I am of the first opinion that on election day a majority will be rolled up over this land of ours for the Republican ticket which will be second only to the overwhelming vote which was given President Warren G. Harding in 1920."

"I am sure that this goodly city of Mansfield and splendid county of Richland will do no less for Calvin Coolidge than they did for Warren G. Harding; I am confident that the results will be all that we expect them to be."

In closing the meeting Chairman Jeffers spoke a few words of approval of the candidacy of Congressman William M. Morgan, Republican candidate for re-election from the seventeenth congressional district, whom he urged be returned to office this fall.

The caravan after a stay in Mansfield of not more than 45 minutes continued on its western route, being scheduled to stop briefly at Bucyrus and then go on to Marion where it was to remain in during the night. The caravan actually stayed in Bucyrus, not Marion as reported by a newspaper.

A string of nearly a dozen automobiles left the Republican headquarters in Bucyrus shortly after 3:00 in the afternoon carrying the welcoming delegation to meet the Coolidge-Dawes caravan at the boundary of Crawford county, enroute from Mansfield. The Bucyrus contingency, along with the caravan, moved through Crestline and Galion and back to Bucyrus.

The group was entertained at the Elks' Club at dinner. The high school band of 60 pieces played at an evening Coolidge-Dawes meeting on the public square at 7:30. General Edward Vollrath will act as chairman and introduce the special guests.

The *Bucyrus Telegraph Forum* reported that "a crowd of 1,000 keeps cool with and without Coolidge as chill breeze sweeps square."

A thousand people shivered in the cold of the Public Square for several hours as they listened to the Republican Caravan Speakers admonishing their hearers to "Keep Kool with Calvin Coolidge."

The Republican notables U.S. Senator Simeon Fess, Congressman R. Clint Cole, Caravan speaker A. W. Jeffries, ex-congressman of Omaha, and H. L. Moore of Plymouth, Vt were introduced by General Edward Vollrath who asked young and old to listen with open heart to the speakers of the evening since political meetings had as their purpose the education of the voter.

Moore gave his usual speech on the Coolidge hometown. Congressman Cole who followed, welcomed the permission of the crowd to keep his hat on, and kept warm rather than cool as he defended the Constitution as it now exists and assailed the amendment offered by LaFollette as tearing away at the foundation of our government, as weakening the super structure and taking from the people the rights and privileges they now have. "There is a movement now a foot to change the three departments of government by weakening the Judiciary, and taking from the Supreme Court the final Jurisdiction on matters of constitutionally and to empower Congress to over ride the court," said Congressman Cole. It is apparent according to Mr. Cole that Congress is now made up of factions, each with a selfish interest, and he asked his hearers whom they believed could best render an impartial decision, Congress or the Supreme Court. If you let Congress decide it might pass laws which would take from you all the rights and priveleges which you now enjoy."

The Democratic party came in for a bit of lampooning and ridicule by a play upon the Bryan Brothers. The difference between W. J. Bryan and Coolidge being, according to Mr. Jeffries, that Bryan likes to talk and Coolidge likes to think. As for Davis, the speaker contended he never heard his name, never saw his picture on the trip. "The poor old Democratic party is just waiting for the day when New York State will give a 1,000,000 majority to Coolidge and Ohio give a half million more."

U. S. Senator Fess, in a husky voice due to a great deal of open air speaking begged to be excused from a long discertation on politics and briefly cited his own record of having always voted the regular Republican ticket.

Great multi-gate campaign card picturing the Republican candidates for Columbiana County, Ohio

Wednesday, September 24
Day 16

Left Bucyrus, Ohio at 8:30 a.m.

Log: Stops at Marion, Kenton, Ada, Lima, and Van Wert, Ohio and Fort Wayne, Indiana. 134 miles. Spent night at Hotel Anthony in Fort Wayne.

Review: Visited the tomb of President Harding and the house his father lived in. Big reception at Ada, by Ohio State Northern University. Senator Fess is president of the university.

Met at the Indiana state line by 300 cars, two bands of music and two aeroplanes.

Big meeting at Fort Wayne baseball park and Red Fire parades thru streets with the Mayor leading on horseback and a fife and drum corps consisting of 24 women.

—E. A. Seidel

Another indication of the tour scheduling and rescheduling as they moved along their journey is apparent in the coverage of *The Marion Daily Star:* "**Auto Caravan Makes Hurried Trip to Marion, Comes Hour Sooner Than Originally Announced**."

As a result of the Coolidge-Dawes Lincoln tour caravan arriving in Marion this morning one hour ahead of the time previously announced, many Marion people were disappointed by failing to see it pass through. The caravan was scheduled to arrive here at 10 o'clock, but word was received late Tuesday evening that it would arrive an hour earlier.

About a dozen machines met the five cars touring to the coast from Plymouth, Vermont, at the Crawford county line and acted as escort to Marion, reaching the court house about two minutes before 9 o'clock. As the clock was striking, the caravan started to move. It went to the cemetery, where a short stop was made at the tomb of President Warren G. Harding. The procession then drove to the Harding home on Mt. Vernon avenue and continued west for Kenton with about 20 machines escorting the party out of the city.

No talks were made here and only a few of the members in the touring party left their machines at the cemetery. It was the aim of the tourists to reach Lima by noon.

Accompanying the tourists through Marion were United States Senator Simeon D. Fess and Congressman R. Clint Cole.

There were four official cars and a truck making up the caravan. The truck contained the picture of the birth place of President Coolidge on each side, while the pictures of Coolidge and Dawes were on the rear.

Many who voted for Abraham Lincoln for president gathered at the courthouse to sign the register for Lincoln voters. Loose leaves from the register were sent to the court house by one of the escorts and signatures were recorded. These will be forwarded to the caravan by mail and become a part of the official register.

Those who signed the register here were Lemuel Pennock, Mt. Victory; Dr. G. T. Harding, Colonel George B. Christian, M. A. Bever, G. B. Willis, A. P. Rhoads, William Herrington, Lucius Carhart, Michael Zachman, Peter Kingery, David Seas, William Brady, Nathaniel Green, George W. Williams, L. Reberr, Elmer St. John, J. H. Buchanan, R. W. Messenger, C. Seiter, J. N. Davids, Michael Click, W. H. Sprague, Schiller Fogleson, W. S. Merrill, T. C. Cram and J. W. Myers.

Strangely, although it was not reported in the articles in the Marion paper, other papers mentioned that there was no meeting held in Marion on account of the universal sorrow in that city following the unexpected death the day before of Gen. Charles E. Sawyer, physician and long time personal friend of President Harding. They reported that the caravan was met at the Marion county line and escorted thru the city by a delegation headed by the county chairman, proceeding directly to the cemetery where they halted for two minutes at the vault where the body of President Harding reposes.

In an article in the *Kenton News Republican* on Sept. 23, the day before the caravan's visit, it outlined plans for the event. "A caravan of Kenton and Hardin county automobiles will proceed to the Hardin-Marion line to meet the cross-continent travelers. Pennants and stickers for the automobiles are being distributed by Attorney Herbert Crane, and will be given out tonight at a meeting in Republican headquarters. These will help carry out the spirit of welcome to the Vermont visitors tomorrow," the article stated.

The September 24 *Kenton News Republican* carried a lengthy front page feature on the event: "**Coolidge-Dawes Tourists Welcomed, Novel Undertaking in History of Politics Meeting Tremendous Ovations; Hundreds of Men and Women here Listen to Talks**."

Hardin county extended to the Coolidge-Dawes Lincoln transcontinental tourists at noon today a welcome, which rivaled the long line of the most remarkable demonstrations in the annals of American politics, which is attending this novel trip across the United States to let the people of America know what his "home town folks" think of President Calvin Coolidge, and his running-mate, General Charles Gates Dawes.

Speakers, with the trans-continental motor tour, who addressed the mass meeting of Kenton and Hardin county citizens from the west steps of the courthouse were United States Senator Simeon D Fess, of Yellow Springs, Ohio, the Hon. A. W. Jeffries, of Omaha, Nebr, and Congressman R. Clint Cole, of Findlay. Attorney Herbert Crane was chairman of the meeting.

Senator Fess contrasted the divided factions of the Democratic party, as exemplified in the New York convention, with the unified spirit of the Republicans in naming Calvin Coolidge as their standard bearer. He referred to the rise of the Progressive party in 1912, in which many of the militant leaders of the Republican party rallied to the support of Theodore Roosevelt. He defied anyone to show him a militant Republican today who is supporting Robert M. LaFollette. He asserted that anyone who is supporting LaFollette and is claiming to be a Republican, is resting upon a shaky Republican record. He recalled that all of the independent newspapers in America are today out and out for Coolidge. "As Maine goes so goes the Nation," he declared. Fess declared that it is generally conceded that New York state will give Coolidge a million majority, and that Ohio will do the same thing.

The participants in the Coolidge-Dawes tour, when the caravan reached Kenton were: Senator S. D. Fess, Yellow Springs, Ohio; Congressman R. C. Cole of Findlay, Ohio; Hon. A. W. Jefferis, of Omaha, Neb., who nominated C. G. Dawes for vice president at the Cleveland convention; John P. Cowan, of Pittsburgh, manager of the tour; Charles T. Doyle, Pittsburgh secretary; A. W. Jefferis, Jr. of Omaha; William Wallace, of Omaha; S. E. Moliden, Brooklyn; Erwin Seldel and E. C. Hoch, of New York; and Herbert L. Moore, of Plymouth, Vt.

When the Coolidge-Dawes Lincoln caravan reached Kenton at 10:30 o'clock this morning, more than an hour ahead of its schedule, the tourists had traveled more than 1,600 miles on their journey from Plymouth, Vt, to the Pacific coast. The early arrival here was explained by the fact that many meetings have been crowded into today's program and in order to keep up with the itinerary, the cars were put in motion at Bucyrus at 8 o'clock.

Jefferies, chairman of the caravan, in addressing the Kenton meeting from the steps of the Hardin county courthouse, was making his 75th speech since leaving Plymouth on the 9th. This included seven states and nearly 80 meetings, not including many roadside stops where the speakers addressed impromptu gatherings of the people.

Those in the official party declared that the caravan had been part of a continuous procession of automobiles all the way from the Green Mountains of Vermont through the Berkshires, up the valley of the Mohawk and across the Catskills in New York, across New Jersey and the entire length of Pennsylvania into the smiling fields of Ohio.

In his address in Kenton Jefferies hit hard on Bryanism:

The American people don't want any more of Bryanism. For more than 30 years the Democratic party has been dominated by W. J. Bryan. Three times the people of the country have repudiated him at the polls. At the New York convention he pointed out that John W. Davis was "John Pierpont Morgan's lawyer" and therefore was undesirable as a candidate. But you know what happened down there in New York? Something slipped and E Davis was nominated. It looked as if Bryan had been defeated. No; nothing of the sort. Davis knew that he must do something to keep Bryan's bitter tongue from wagging, and to make peace, he offered the nomination of the vice presidency to "Brother Charley." Of course that pleased W. J. and the old team work of the brothers goes merrily on. Nobody expects Davis to be elected but they are maneuvering with the LaFollette crowd to create a deadlock in the electoral college in the hope that the Senate will name Bryan vice president and thus, have him promoted on March 4 next to the presidency. But, good friends, of Kenton, don't fool yourselves. The people of Ohio don't want Bryan; nobody in his sober senses wants Bryan. And we won't have Bryan. The sentiment we have heard in our journey of nearly 2,000 miles convinces us that the people of the United States have their feet on the ground in this campaign and that there will be a sweeping victory for Coolidge and Dawes.

The tourists held a luncheon meeting in Kenton and then moved on to Ada where they made a brief address to the students and residents of the town.

Thirty automobiles of Allen County Republican leaders greeted the Coolidge-Dawes Lincoln tourists shortly after 1 p.m. near Ada and escorted the entourage into Lima. A large mass meeting was held before the home of Dr. A. S. Rudy, 215 West North Street with Congressman John L. Cable as meeting chairman. Senator Fess and Allen County candidates spoke at the rally.

"Big Jeff" addressing audience in Ada, Ohio

Towns along the caravan route did not have a great deal of lead time in planning for the event. The auto caravan passed through Lima, Ohio on September 24. The *Lima Republican Gazette* reported on September 5 that according to information received that morning by H. E. Garling, chairman of the Allen county Republican executive committee Lima would be visited by the motor tourists. Two days later an article relates "**G.O.P. Leaders Plan Coolidge Tour Welcome, Big Turnout and Luncheon To Be Arranged For Political Tourists**." By September 13, Captain Frank Wiach, big game hunter and old friend of Theodore Roosevelt and Buffalo Bill, and Samuel W. Long, magazine writer, acting in tandem as volunteer advance agents for the Home Town Coolidge Club, visited Lima to make arrangements for the visit to the city. They provided hotel, garage and medical facilities for the caravan. After all the planning, the caravan didn't get to Lima until 2 p.m., having had a luncheon meeting in Kenton.

The *Lima Republican Gazette* reported on the caravan's visit:

Coolidge Dawes cloth banner. 44" x 29"

G.O.P. Caravan Party in Lima. Lima Republicans Entertain Home Town Club; Cong. Cable, Sen. Fess Speak

Lima greeted the Coolidge-Dawes Lincoln tourists Wednesday. Shortly after one p.m., over 30 automobiles containing the Republican leaders of the county left for the Ada road, where they met the caravan at 2 p.m.

Plans called for a large mass meeting before the home of Dr. A. S. Rudy, 215 West North street. Congressman John L. Cable was chosen chairman for the meeting, which was to be addressed by members of the Home Town Coolidge Club, Senator Simeon D. Fess, and Allen county candidates.

The tourists held a luncheon meeting in Kenton, and came from there through Ada to Lima.

At each stop the caravan opens the Lincoln Register for the signature of all old voters who cast a ballet for Abraham Lincoln. the number who have already signed the document far exceeds the expectations of the tour manager. The veteran Republicans are given a place of honor at each meeting.

It is planned to present this historical volume to President Coolidge after election day. He probably will place it in the archives of the government.

"Keep Cool-idge" felt campaign pennants. 31" long

Thursday, September 25
Day 17

Left Fort Wayne, Indiana at 8:30 a.m.

Log: Stops at Ligonier, Goshen, Elkhart and South Bend, Indiana. 108 miles. Spent night at Hotel Oliver, South Bend.

Review: Aeroplanes flew with us all day, dropping hand-dodgers in all the towns.

This is the best organized state we have been into and all stops are being made according to schedule.

Lincoln now being driven by J. A. Smith of Chicago.

Joined here tonight be newspapermen from Chicago and Kansas City. Senator Fess of Ohio is leaving us tonight in South Bend. This town is the home of the famous Notre Dame College.

Big parade and large outdoor meeting held on steps of the Court House.

—E. A. Seidel

The September 25 *Fort Wayne Journal Gazette* covered the G.O.P. rally:

CARAVAN SPEAKER RAPS THIRD PARTY

Senator Fess Says America Will Never Russianize Herself

Between 2,500 and 3,000 people from all parts of the Twelfth district attended the Republican rally held last night at the Lincoln Life baseball park.

The rally was a reception to the Coolidge-Dawes caravan which is crossing the continent in behalf of the Republican nominees for president and vice-president. It started from Plymouth, Vt., the birthplace of President Coolidge and is following the Lincoln highway.

A street parade in which 58 automobiles took part was held prior to the speaking. Owing to the fact that the caravan did not arrive in the city until almost 7 o'clock the speaking program did not begin until 9 o'clock.

Senator George Wharton Pepper, of Pennsylvania, who had been announced as the principal speaker, was not with the caravan and the address of the evening was delivered by Simeon Fess, United States senator from Ohio.

Senator Fess steered straight away from a discussion of the issues of the campaign, except as to foreign policy, and talked about the personality of the candidates. He appealed to the audience to elect Coolidge and Dawes, a team that has never been matched in American history.

The speaker took a rap at the third party movement and declared that we will never Russianize America under the leadership of Bob LaFollette or any other leader.

Senator Fess declared that America would never enter the League of Nations. The largest part of his address was consumed with a discussion and praise of the work of the reparations commission. He declared that the foreign policy of the United States under the Harding-Coolidge administration has been the most brilliant foreign policy it has ever had.

There never was a time, the speaker said, when so many Democrats will vote the Republican ticket as in 1924. That is not because Mr. Davis is not a very capable and very competent man. I am somewhat distressed over the character of the speeches that Mr. Davis has been making. They do not sound like the man I know so well and so favorable. He says the nation has no foreign policy. That is strange when it never had such a brilliant foreign policy as it has today.

Senator Fess closed by discussing the possibility that LaFollette might get enough votes to throw the election into Congress. In such a case, he said, it would be impossible for the house to elect a president and that Charles W. Bryan, Democratic candidate for vice-president, has "nine chances out of ten" to become president, if the election goes to the house.

It is not going to happen, but it might, he said, suppose enough people should be deluded by the third party movement, which I know they will not. There is only one way to stop it and that is to make the Coolidge majority so big that there can be no doubt.

The temperature was low in the ball park and at the conclusion of the address by Senator Fess nearly two-thirds of the audience went home."

Other speakers included boyhood friend of Coolidge, Herbert Moore, Congressman A. W. Jeffries of Nebraska, Will R. Wood, Republican congressman from the Tenth Indiana district, William McIntosh, who said he voted twice for Abraham Lincoln, and Attorney Clyde Carlin of Angola.

The Indiana State Committee and members of the Caravan pose in front of the Speed Wagon in South Bend

The *Goshen Daily News-Times* for September 25 gives the Coolidge-Dawes Caravan front page top headlines:

COOLIDGE-DAWES CARAVAN HERE TODAY

Senator Fess of Ohio and Others Speak.
Transcontinental Party Stops in Goshen for Lunch This Noon.
Crowd of Over Thousand Gathers in Courthouse Park to Hear Opening Address of National Campaign.

The Coolidge-Dawes transcontinental caravan, following the route of the Lincoln Highway from New York to San Francisco, arrived in Goshen at 12:30 this noon, just a few minutes behind schedule. A meeting was held in the court house park where a crowd of more than a thousand persons had gathered. Herbert Moore, boyhood friend of President Calvin Coolidge, Senator Simeon D. Fess of Ohio, and Congressman Will R. Wood of Lafayette, Ind., were the speakers here.

The caravan was greeted at the Elkhart-Noble county line by a delegation of Elkhart county republicans who filled more than a hundred automobiles. The cars were parked for a long distance along the highway, County Chairman D. M. Hoover occupying the last car in the meet.

When the travelers arrived, accompanied by a Noble county delegation, the Noble county republican chairman presented the Indiana state flag to Chairman Hoover, who carried it across Elkhart county. The chairman and the official caravan then headed the procession and led into the city.

Introduces Speakers

Aldo Simpson, of Goshen, republican vice chairman for Elkhart county, introduced Dr. M. M. King, of Indianapolis who is with the party as the representative of State Chairman Clyde Walb. Dr. King introduced the speakers of the party. First was Herbert Moore, of Plymouth, Vermont, who owns an adjoining farm to that of President Coolidge's father, and who was a boyhood friend of the president. Moore spoke of the president's boyhood life, and how the man who today occupies the highest position of honor in the country made the most of his slender opportunities.

Senator Simeon D. Fess of Ohio, was the next speaker, and delivered a forceful address which was received with much enthusiasm. Will R. Wood, a congressman from Lafayette, Ind., was the next to speak, making a plea for the election of Congressman Andrew J. Hickey, from this district. Former Congressman W. A. Jeffries of Nebraska and Rep. Andrew J. Hickey of LaPorte also spoke.

The Pan-American quartet, composed of saxophone players from Elkhart, provided music on the court house steps, entertaining the crowd until the arrival of the caravan. Members of the party took turns in going to lunch at the Alderman, while the meeting was being held here.

In the party were also Arthur Gilliom, of South Bend, candidate for attorney general, Bernhard Urbahna, candidate for state treasurer, and Judge Willoughby of this supreme court, candidate for reelection. Ed Jackson, republican candidate for governor, was unable to be here.

The caravan left Goshen shortly before two o'clock, going to Elkhart, where another meeting was held this afternoon. The caravan was composed of six official automobiles carrying the speakers and a truck carrying a miniature of the president's home in the Vermont hills.

Another stop was also scheduled for Mishawaka, and the caravan was to stop in South Bend for the night. A big meeting is planned for this evening in Howard Park, South Bend.

The South Bend Tribune, September 26 edition gives the Caravan the top headline on the front page and photos inside.

START G.O.P. DRIVE HERE.

Republicanism is Expounded to Big Crowd. Coolidge-Dawes Caravan speakers Heard During Local Stop.

Confident that Calvin Coolidge will be re-elected president of the United States, Nov. 4, but pointing out the danger of inactivity on the part of voters favorably inclined to the republican ticket, speakers attending the Coolidge-Dawes caravan on its stop here, addressed a crowd of several thousand people at the courthouse corner last evening, official opening the republican speaking campaign in South Bend.

Both LaFollette and Davis were scored by several speakers. The principle addresses were made by Congressman A. W. Jefferis of Omaha, Neb., Congressman Will Wood of the 10th Indiana district and Herbert Moore of Plymouth, Vt, a boyhood friend of President Coolidge. State Republican Chairman Clyde L. Walb, of LaGrange, Ind, accompanied the caravan in its sweep through Indiana. The caravan was met by large and enthusiastic crowds at every stop. It proceeded this morning on its way to the west coast with stops arranged at Gary, Ind., and elsewhere in the Calumet district, before proceeding into Chicago.

Walb is Confident

While the republican open-air meeting was being held, the independent speaking campaign in south Bend was opened at the Central Labor union hall, 315 S. Michigan street, with an address by James McGill, of Valparaiso, Ind, who spoke for the candidates of LaFollette and Wheeler. About 35 were at the meeting. The democratic speaking campaign was opened last week with an address here by John W. Davis, nominee for president.

State Chairman Walb, in a statement to the Tribune, said: "Coolidge and Dawes are strongly the favorites in Indiana and I have full confidence that the state will go out and tell the country the facts about the president."

The rest of the article basically covers the speeches of Jefferies and Herb Moore.

Friday, September 26
Day 18

Left South Bend, Indiana at 8:00 a.m.

Log: Stops at LaPorte, Valpariso, Gary, East Chicago, Chicago Heights and Joliet, Illinois. 159 miles. Spent night at Park Hotel in Joliet.

Review: Flag presentation at East Chicago.

Senator M. W. Pitt of Iowa joined us at Gary, Indiana. Long drive today, and did not finish until 10 o'clock.

Suggested by Jeff, Jr. that Loeb and Leopold come out and meet us with the prison band at Joliet.

—E. A. Seidel

The Chicago Heights Star carried a front page story:

COOLIDGE-DAWES CAVALCADE ARRIVES.

Comes in From Blue Island Because of Change of Schedule. 91 Cars Leave Here for Hammond to Escort the Visitors – Incidents of the Journey

Now in the eastern half of the country, the Coolidge-Dawes Lincoln tour has become history, and in a short time it will be history in the whole country, and we will know whether or not it is made in vain. Chicago Heights had a fleeting glimpse Friday night of the tourists, and of the "covered wagon," which contained a picture of the home of President Coolidge. Not because the people here and in surrounding territory would not have been pleased with more than a fleeting glimpse. They were on hand in large numbers early in the day, and a great number of cars were ready to go to Dyer, Ind., and escort the distinguished visitors making up the entourage to the city, while thousands were anxious to hear what the speakers had to say.

Much Enthusiasm

Congressman Elliot W. Sproul, who had charge of the tour in his congressional district, had managed to arouse much enthusiasm over the coming of the visitors, and was here at the appointed time, ready to lead the hosts to Dyer to meet the tourists. Everybody was rearing to go, and Chief Cornilsen had a squad of his finest, with Officer Baker as motor cycle man, to act as orderly.

But it appeared that somebody had monkeyed with the arrangements, and the tour was somewhat "balled up." The Indiana State Automobile association had taken charge of the visitors when they hit Hoosier territory and had run them back and forward over that state, and early in the afternoon came information that the visitors would not come by way of Dyer, but Blue Island, and Chicago Heights, and thence on their westward sweep. They were due in Joliet at 8 p.m., and were behind schedule, but the fact didn't daunt the Republicans here who had prepared to escort the visitors.

Hammond Unprepared

It was stated that the tourists would reach Hammond about 3 p.m., and the local crowd, in 91 automobiles, all properly decorated with "Coolidge and Dawes" and Chicago Heights banners, set out for Hammond. They found the town oblivious of the coming of the visitors, with nary a flag or decoration visible, and only the Chicago Heights contingent to welcome the visitors when they arrived. Later they were joined by a big delegation from Calumet City (formerly West Hammond), and all gathered about the public library square and waited until nearly 7 o'clock before the visitors arrived.

When they pulled in, escorted by some of the Indiana autoists, and Congressman William R. Wood, of Lafayette, a large number of those who started on the tour had become discouraged, and left for home. However, the program was carried out. Congressman Wood introduced Hon. Albert W. Jefferies, former congressman from the second district of Nebraska, and the man who nominated Charles G. Dawes for vice president, and who notified Mr. Dawes at his Evanston home, officially, of his nomination. Mr. Jefferies is an incisive speaker, but in a five or ten minute speech he scarcely gets warmed up. Even in that time, however, he took some vigorous whacks at the Democrats, and their doings, and "razzed" Mr. LaFollette and his "Red" associates, who would mangle the constitution, establish government ownership of railroads, and once in control of the government would emulate the Russians now in power, and perpetuate their rule.

Will Visit Seventeen States

From Hammond the cavalcade moved to Harvey, with banners flying, and halted at the park opposite the Elks' club, where Congressman Sproul introduced Mr. Jefferies to the people. Quite a crowd had assembled, and Mr. Jefferies continued his lambasting of the Democrats. Then the road was resumed, and the party journeyed to Blue Island, but because of the fact that they were so far behind schedule, they did not stop, but waved greetings to the people as they sped along.

Chicago Heights was reached at 8 p.m. Word had been passed that the speaking would be in the forest preserve at Fourteenth street and Chicago road, and a large crowd was on hand to listen to the speakers. Congressman Sproul again presented Mr. Jefferies, who spoke briefly in the vein of his previous speeches, and then the cavalcade resumed its journey.

This cross country tour was conceived by the boyhood friends of President Coolidge, "for the primary purpose of acquainting the voters of the country with the Calvin Coolidge they know." It will traverse seventeen states before it finishes, making most of the journey over the Lincoln highway, and visiting 300 cities and towns on the way. From San Francisco, the tourists will go north through Oregon and Washington to Seattle and Bellingham.

Saturday, September 27
Day 19

Left Joliet, Illinois at 9:00 a.m.

Log: Stops at Aurora, DeKalb, Rochelle, Dixon and Sterling, Illinois. 109 miles. Spent night at Lincoln Tavern, Sterling.

Review: Rained all day yet people turned out to see us everywhere.

Meeting held in Courthouse at Dixon, and a large enthusiastic audience greeted us in the Academy of Music in Sterling.

The Caravan is resting here over Sunday.

—E. A. Seidel

The Aurora Daily Journal reported: **"Coolidge-Dawes Caravan Here En Route West; Escorted Thru City by 50 Automobiles."**

The Coolidge-Dawes automobile caravan, en route from Plymouth, Vt., to the western coast, via the Lincoln highway arrived in Aurora this morning shortly after 10 o'clock and paraded thru the downtown streets led by the St. Charles School for boys band and escorted by 50 automobiles from Elgin and Aurora.

The local automobile escort met the caravan at Plainfield. Following the parade the members of the party gathered at Lincoln park where brief addresses were given.

The caravan was presented an Illinois flag by Mrs. J. H. Bliss, who was in charge of the women in the reception party. The caravan included six automobiles, one of which represented the log cabin birthplace of Coolidge. The delegation left Plymouth, September 9, and expects to be in San Francisco by October 20. The machines have covered more than 1,200 miles passing thru New Hampshire, New York, New Jersey, Pennsylvania, Ohio, Indiana and Illinois. A meeting, the 125th of the trip, was held last night in Joliet.

A list of all the men who voted for Lincoln was taken along the route. Three Aurorans signed their names to the roll. They were: Capt. J. F. Harral, Capt. J. H. Freeman and William E. Holmes.

Herbert Moore was given extensive coverage in the Aurora paper's account. Besides his vivid description of Plymouth as a simple community and his antedotes about growing up with Cal, he talked about the Coolidge home inaugural and his sons.

The town is 12 miles from the railroad. No electic lights near there. There are two phones and Cal's father has not either. He has never had a phone because he has an old maid housekeeper who refuses to answer one. When the great catastrophe to the nation came, and news of President Harding's death was scorching the wires, the telephones in Plymouth rang and no one heard them. A man in a car brought the news and soon Washington notified Coolidge that he must be sworn in as president at once. Did he call a supreme court judge to do it? No. He asked his father if he were still a justice of the peace and was sworn in then and there without pomp. A simple man.

While he was president, his sons were working, one in the tobacco fields and the other in a military camp. They kept right on working just as he is doing today down in Washington while his campaign is going on. The reason we started this campaign was because LaFollette said that Coolidge was nothing but a cold blooded aristocrat and had never seen a farm. We in his home town got together, the 29 of us, and with a friend in Pittsburgh started out to tell the people facts and that's all we need in this campaign, facts.

DeKalb, Illinois - Waiting for rally to start

THE 1924 PRESIDENTIAL CANDIDATES

A brief review of the lives of the presidential candidates of 1924 quickly reflects that they did not have a great deal in common.

Calvin Coolidge – Republican Presidential Candidate

Born in Plymouth, Vermont, on July 4, 1872, Calvin Coolidge was of frugal Yankee stock. Graduating from Amherst College, Coolidge soon became involved in politics, moving up the political ladder from councilman in Northampton through several political offices, to Governor of Massachusetts, and then selected as the vice presidential candidate in 1920 for the Harding-Coolidge Republican ticket. At 2:30 on the morning of August 3, 1923, while visiting in Vermont, Calvin Coolidge received word that Harding had died and he was President. By the light of a kerosene lamp, John Coolidge, Calvin's father, who was a notary public, administered the oath of office. Coolidge was "distinguished for character more than for heroic achievement," wrote a Democratic admirer, Alfred E. Smith. "His great task was to restore the dignity and prestige of the Presidency when it had reached the lowest ebb in our history...in a time of extravagance and waste." Known as "Silent Cal," the President was a man of few words and perfectly satisfied to maintain the status quo in a period of contentment in the country. Although many saw him as remote, he was one of the most accessible Presidents, constantly greeting a variety of delegations to the White House and being the first President to make strong use of the radio to convey his messages. Both his dry Yankee wit and his frugality with words became legendary. After leaving the Presidency in 1929, Coolidge wrote a series of syndicated daily newspaper columns. He died in January 1933 and is buried in the family plot at the cemetery in Plymouth.

Charles Gates Dawes – Republican Vice Presidential Candidate

Charles Gates Dawes was born in Marietta, Ohio, in 1865 and graduated from Marietta College and the Cincinnati Law School. Dawes practiced law in Lincoln, Nebraska, from 1887 to 1894 and then became interested in public utilities and banking. He became Comptroller of the Currency, U.S. Treasury Department, from 1898-1901 in the McKinley Administration, and was an unsuccessful candidate for the United States Senate in 1902. Dawes became a brigadier general during the First World War and in 1921 because the first director of the Liquidation Commission in the War Department. Upon the creation of the Bureau of the Budget,

Dawes resigned his War Department position and became the new Bureau's first director in 1921. He worked on a program to enable Germany to restore and stabilize its economy as part of the Allied Reparations Commission in 1923, which eventually resulted in his winning a Nobel Peace Prize in 1925. Chosen as Coolidge's running mate at the convention in 1924, the Coolidge-Dawes ticket went on to win, however the two never really were a team, or even friends. Dawes became Ambassador to Great Britain from 1929-1932 during the Hoover Administration and then resumed the banking business as chairman of the board of the City National Bank & Trust Company in Chicago from 1932 until his death in April 1951 at the age of 85. Dawes was a disciplined and productive man. In addition to being a banker, a general, a Nobel Prize winner and vice President, Dawes wrote nine books and was involved in countless civic duties. Even in music he excelled. He performed on the flute and piano and his 1912 composition "Melody in A Major" became a pop song ("It's All in the Game") in 1951 and was recorded by Van Morrison, Cliff Richard, Barry Manilow and Elton John.

John W. Davis – Democratic Presidential Candidate

John W. Davis was born in 1873 in Clarksburg, West Virginia. He acquired his father's conservative politics and staunch allegiance to the Democratic Party, even as he later represented the interest of conservative business interests opposed to the New Deal of FDR. Davis graduated with a law degree from Washington and Lee University and represented West Virginia in the U.S. House of Representatives from 1911 to 1913. He was an ambassador to the United Kingdom from 1918 to 1921 in the Wilson Administration. In 1922 Davis rejected appointment to the U.S. Supreme Court. Davis was a dark horse candidate for the Democratic nomination in both 1920 and 1924, winning the 1924 nomination as a compromise candidate on the 103rd ballot. His denunciation of the Ku Klux Klan and his prior defense of black voting rights as Solicitor General under Wilson cost him votes in the South and among conservative Democrats elsewhere. Davis was one of the most prominent and successful lawyers of the first half of the twentieth century, arguing 140 cases before the U.S. Supreme Court, more than anyone had argued to that time. He may be best remembered for successfully defending the steel industry against government seizure during the Korean War, and for unsuccessfully arguing South Carolina's case for maintaining segregated schools in the school desegregation cases now known jointly as Brown v. Board of Education. Davis died on March 24, 1955, just short of his 82nd birthday.

Charles W. Bryan – Democratic Vice Presidential Candidate

Charles Wayland Bryan was born on February 10, 1867 in Salem, Illinois, the younger brother of perennial U.S. Democratic presidential candidate William Jennings Bryan. Bryan served as mayor of Lincoln, Nebraska, from 1915 to 1917, and again from 1935 to 1937, and as Governor of Nebraska from 1923 to 1925 and again from 1931 to 1935. He was an unsuccessful candidate for governor in 1926, 1928 and 1938. Bryan was selected as the Democratic vice presidential candidate in 1924, where he was picked largely because of his name association with his brother, to serve as running mate to conservative easterner John W. Davis. He died in 1945 in Lincoln where he is buried.

Robert M. La Follette – Progressive Presidential Candidate

Robert Marion La Follette, Sr. was born on June 14, 1855 in Primrose, Wisconsin, just outside of Madison. In 1879, La Follette graduated from the University of Wisconsin-Madison after studying law. La Follette was elected Dane County District Attorney in 1880 and four years later he won election to the United States House of Representatives where he served until 1890. He became the 20th Governor of Wisconsin from 1901-1906, after two unsuccessful attempts, and U.S. Senator from Wisconsin from 1905-1925 as a member of the Republican Party. Both as governor and senator, La Follette championed progressive causes. A brilliant orator given to periodic bouts of "nerves," he made many enemies over the years, particularly for his opposition to the United States' entry into World War I and his defense of freedom of speech during wartime. While in the Senate, besides strongly opposing America's involvement in World War I, he campaigned for child labor laws, social security, women's suffrage, and other progressive reforms. In 1912, La Follette ran for the Republican Party nomination for President of the United States, but lost to William Howard Taft, due largely to many Progressives supporting Theodore Roosevelt's third-party candidacy. He bitterly opposed Roosevelt, blaming him for sabotaging his campaign. He backed Wilson over Taft and Roosevelt in that year's election. In 1924 La Follette, although nominally a Republican, again ran for President, this time on his own Progressive Party ticket, coming in third with 17% of the popular vote and carried Wisconsin. He died several months later. He is best remembered as an exponent of Progressivism and for fighting bossism. In 1957, a committee led by Senator John F. Kennedy selected La Follette as one of five of their greatest Senate predecessors whose oval portrait would adorn the President's Room off the Senate floor.

Burton K. Wheeler – Progressive Vice Presidential Candidate

Burton Kendall Wheeler was born in Hudson, Massachusetts, on February 27, 1882. He graduated from the law department at the University of Michigan in 1905, was admitted to the bar the same year and commenced practice in Butte, Montana. In 1910 Wheeler started his political career as a member of the State house of representatives. He was an unsuccessful candidate for governor of Montana in 1920, but elected as a Democrat to the United States Senate in 1922. Wheeler, a young Democratic senator was selected as La Follette's running mate in 1924 at the Progressive Party's convention in Cleveland, Ohio. Wheeler was reelected to the United States Senate as a Democrat in 1928, 1934 and 1940. In 1940, there was a large movement to "Draft Wheeler" into the presidential race, possibly as a third party candidate, led primarily by labor's John L. Lewis. As tensions mounted in Europe, Wheeler became known for his isolationist beliefs. After the beginning of World War II in Europe, he continued to oppose any aid to Britain or any other countries involved in the war, becoming a supporter of the American First Committee. Wheeler was an unsuccessful candidate for renomination to the Senate in 1946 and returned to his practice of law. Burton Wheeler died in Washington, D.C. in January 1975 just short of his 93rd birthday.

Sunday, September 28
Day 20

Lay over at Sterling, Illinois

The *Sterling Daily Gazette* for September 27 carries the article:

CARAVAN STOPS OVER NIGHT HERE.

Led by Float Bearing Model of Coolidge Home, Caravan Is Enroute.

Led by the great float bearing a model of the Vermont Home of Calvin Coolidge, the Coolidge-Dawes caravan rolled down the Lincoln highway today.

A patriotic demonstration and speech at Aurora, Ill., started the day's program at 10 a.m. Then the caravan, joined by scores of automobiles, driven by enthusiastic republicans, proceeded to DeKalb, and thence to Sterling.

The caravan will rest tonight and Sunday at Sterling. It will proceed, according to schedule, to Cedar Rapids, Ia., Monday, for a Monday night meeting. Tuesday it will roll on to Marshalltown for a similar evening meeting. Thursday the troop will drive to Omaha, bound for the coast.

Coolidge-Dawes Caravan rolls westward on the Lincoln Highway

In the Wednesday, September 24 edition of the *Sterling Daily Gazette* there was an article about the plans for the caravan and an invitation extended to General Dawes.

BIG PARADE TO PRECEDE BIG G.O.P. MEETING.

Hope To Have Chas. G. Dawes, Republican Candidate for Vice President as Principal Speaker

The republican club, of this city has about completed all plans for the arrival here on next Saturday evening of the Coolidge-Dawes Lincoln Highway caravan. Without doubt the meeting Saturday evening will be the greatest republican meeting ever held in this county in years. Political leaders in a number of cities surrounding Sterling have stated that large delegations will be here to greet the caravan.

Arrangements have been completed for the parade which will start promptly at 7:30 p.m. at the Lincoln Tavern, and if the weather is favorable the line of march will be west on Fourth street, to the Elks, south to third street thence east to Third Avenue and to the park. If the weather is unfavorable for an outdoor meeting, the parade will turn off on First avenue north to Fourth street and then to the Academy of Music.

A list of the speakers, who will be here, will be published Thursday. The speaker's bureau is endeavoring to secure some of the leading speakers for this meeting, realizing that it is to be one of considerable magnitude.

General Dawes has been invited to attend and the committee is awaiting a reply to their message urging his attendance here on that evening. At least a hundred local cars will motor to Dixon Saturday afternoon to meet the caravan. They will leave the Buick garage on Locust street at two o'clock.

Illinois Coolidge bumper sticker (13")

Monday, September 29
Day 21

Left Sterling, Illinois at 8:30 a.m.

Log: Stops at Morrison, Mississippi River Bridge, Clinton, Iowa, DeWitt and Cedar Rapids, Iowa. 113 miles. Spent night at Hotel Montrose in Cedar Rapids.

Review: Iowa flag presentation at Clinton, Iowa. L. L. Hasbrouck of Des Moines joined us at Clinton to drive the Lincoln.

An additional speaker was added at Cedar Rapids in the person of Major Dent Atkinson, the world traveller and lecturer.

The Rotary Club at De Witt entertained us at luncheon, also the Women's Republican Club. Big parade and meeting in Cedar Rapids.

—E. A. Seidel

The Cedar Rapids newspapers payed tribute to the caravan's visit. One accounting:

Big Reception for Coolidge-Dawes Troupe. Two Hundred Cars In Parade When G.O.P. Caravan Reaches This City.

Escorted down First Avenue by more than 200 automobiles, the Coolidge-Dawes transcontinental caravan rolled into Cedar Rapids, flags flying and banners blowing in the breeze, promptly on schedule at 7 p.m. yesterday.

Headed by the band of Hanford post, American Legion, city officials, Civil War veterans, and local republican leaders, it moved between lines of hundreds of cars parked at the curbings and intersections which added the din of sirens to the welcoming music of the band and buglers. Hundreds standing on the sidewalks and in the streets, scores who presented their faces against windows of homes, just lighted for the evening, or rushed to opened doors, added their bit to the enthusiastic welcome given by 800 or more who occupied the escort cars.

Cedar Rapids meant more than just another vociferous greeting, another nights bivouac for the weary, though enthusiastic trekkers. It marked the half way point in distance and in time on their cross-continent trip. They have come approximately 2,500 miles since leaving Plymouth, Vt, President Coolidge's birthplace, Sept. 9. This morning they wheeled westward again on the last half of their journey to San Francisco, where the caravan is due Oct. 20. Though that is their definite objective, plans now call for a swing northward along the Pacific coast to Seattle.

Biggest Welcome Here

The welcome in Cedar Rapids was characteristic – though on a larger scale of course, of the reception given the caravan at every town along the Lincoln highway, since it entered the Hawkeye state yesterday forenoon at Clinton. From 150 to 300 persons gathered around the six automobiles comprising the caravan at every little town, listening eagerly and applauding frequently the four-minute speeches of A. W. Jefferies, Omaha, director of the tour; Herbert L. Moore, Vermont farmer and boyhood schoolmate of President Coolidge, and Major Dent Atkinson, world-traveler and lecturer.

Flags decorated the main streets of the towns on the caravan's itinerary. Pioneers who voted for Abraham Lincoln, the first republican President, proudly wrote their names in the register book of the caravan. Folk stopped their work, merchants left their stores and shops, farmers came to town to see and hear. The caravan reached Stanwood just after the public school was dismissed and scores of school children lined the sidewalks. The high school football team, sure not to miss the caravan, ran signals in the main street until it arrived. There too, Martin Johnson and John Doudney who voted for Lincoln, signed the records.

The caravan became to these Iowa folk as it has to thousands in states through which it has passed, the voice of the republican party, telling in simple and understandable terms who its candidates are and what they stand for. The caravan is something unusual in political campaigning. It is new, interesting and decidedly picturesque. It grip's the public's attention and holds the individual's interest. This was apparent at every town. It is fulfilling its purpose – reaching folk in communities not ordinarily marked on the itinerary of political speakers. It is doing what Herb Moore says it was primarily organized for – to refute the reported statement of LaFollette that "Calvin Coolidge is a cold-blooded aristocrat and never saw a farm." the people of Mr. Coolidge's home town were indignant. They wanted to tell the world, LaFollette was wrong.

Portraits on Truck

So they are unrolling a ribbon of Coolidge-Dawes enthusiasm across the continent. On the truck, which is the car of central interest from the spectator's standpoint are pictures of President Coolidge's birthplace at Plymouth, Vt. But that isn't all. On the rear doors of this new covered wagon are painted the portraits of Calvin Coolidge and Charles G. Dawes and the words, "Two common sense Americans."

On the truck's cab, plain at night as well as by day is the slogan "Common Sense and Brass Tacks."

This truck carries the caravan's baggage. Incidentally, it is piloted by E. A. Seidel, son of H. Seidel, sales manager of the Reo Automobile company.

Cedar Rapids did not wait for the caravan to come to it. It went out to meet it. Rep. Cyrenus Cole, S. G. Armstrong and a newspaper man in Mr. Armstrong's car, met the travelers at Lowden. Mr. Cole spoke there and also at Stanwood, in addition to the caravan's spellbinders.

Just east of Stanwood, Harrison E. Spangler, Linn county republican chairman, A. S. Reed, Will Kurtz of Lisbon, H. M. Wislon, American legion bugler in uniform, and an Evening Gazette representative, in Mr. Reed's car, met the caravan as the official pilot car into Linn county and Cedar Rapids.

Short stops were made in Stanwood, Mechanicsville, Lisbon, Mt. Vernon and Marion. At the latter place, the last before entering Cedar Rapids, 25 automobiles headed by Mayor A. E. Crewe, James Bowman, County Auditor G. A. Senninger, Dr. J. J. Booth and Judge and Mrs. F. L. Anderson, were in line just outside of town and escorted the campaigners into Marion.

Receive Welcome

Spurred on by the desire to maintain its schedule, hurried by the pangs of hunger and chill September wind, the caravan had picked up speed between Mount Vernon and Marion and it was a few minutes ahead of time when David Turner, huddled in the side car of a police motorcycle driven by Gordon Hughes, shouted "hello" and "we're ready for you" just east of Kenwood Park.

The sharp notes of Wilson's bugle foretold the approach of the cavalcade. A myriad of lights, in response, told the tourists of the line-up waiting at Twenty-seventh street and First avenue.

There Ray Murrel, chairman of the reception committee, Frank K. Hahn, chairman of the parade, and their assistants were ready with their part. There was but a moment's pause while the local cars which were to lead got under way and then the caravan wheeled smoothly to the half way mark of its trail across the continent.

There are fifteen in all in the caravan's party. John P. Cohan, Pittsburgh newspaper man, from whose mind the basic ideas for the caravan came, is the manager. Charles T. Doyle, also of Pittsburgh, is the secretary. Drivers who are also their own mechanics and their assistants, complete the personnel.

Eight states have already been traversed by the political pilgrimage – New Hampshire, Massachusetts, New York, New Jersey, Pennsylvania, Ohio, Indiana, Illinois.

There have been as many as 5,000 automobiles, four abreast, in the procession escorting it, a parade fifteen miles long, Mr. Cowan said. Through Pennsylvania and Indiana there was an almost continuous line of cars as one town took up the escort where another left it. The enthusiasm, everywhere, especially in the middle west, belies the rumors of LaFollette strongholds or democratic gains, and predicts republican victory, he declares.

DeWitt, Iowa Women's Coolidge Club (note pennants)

A second Cedar Rapids newspaper account:

Coolidge Much Praised by Caravaners. Big Crowd Hears Talk at Auditorium Made By Members of Home-Town Club. LaFollette Was Given Many Knocks. Speeches All Greeted By Hearty Applause From Crowd Jamming Building.

Battling against frequent volleys of applause that all but stopped their orations, three close friends of Calvin Coolidge, all members of the Coolidge-Dawes Home-Town Club caravan, last night espoused the cause of republicanism and the virtues of Calvin Coolidge before a crowd of Cedar Rapids citizens that packed the city auditorium to its very doors. Conservative estimates placed the crowd at 2,000 persons. Other figures ranged much higher.

The meeting was opened by Congressman Cyrenus Cole who presided and introduced the three speakers. The congressman reviewed the series of glorious receptions that had been tendered the caravan since its arrival in Iowa and pointed to the speakers as the exponents of the spirit shown by the Iowa people in their receptions to the caravan.

Major Dent Atkinson of New York City, the first speaker on the program, kept the crowd in an almost continual uproar with his British wit followed in each instance by volleys of applause as he recounted the record of the present administration.

Best Place To Live

I have been a wanderer all my life. I have traveled in almost every country on the globe and am now reeling off my second 2,000,000 miles, Atkinson said, but never have I seen in all my travels a country so great and so full of opportunities as the United States. I am an American by choice – by adoption but if I had that choice to make again, if I had the choice of all the countries on the globe in which to make my residence, I would choose the United States without a moment's hesitation, declared Atkinson.

I have been wounded twice defending the stars and stripes and I am willing to be wounded again and again for the cause but I would not shed one little drop of my mother-in-law's blood for any other. Do not let any scheming politicians fool you about the advantages of Russia or any other nation other than the United States. The rats are eating more in America, are better fed in Iowa, than the human beings are in Europe, yet some persons want to change conditions here to match those of Europe.

LaFollette Un-American

LaFollette, elected under the flag of the republican party, lives in a $200,000 home and draws a $7,500 salary from the government, thanks to the party that put him in the senate and yet he is biting the hand that feeds him. He is decidedly un-American. He is of the reds, the same brands of reds that burned our farms and our crops in Montana and of whom we had to shoot down eight members to stop their dastardly practices. Do you want men like that in the Presidency of the United States?

The other day I visited at Oyster Bay, the home of that most illustrious American, Theodore Roosevelt, to the soul of whom LaFollette seeks to hitch his wagon. He can never do it in a million years because Roosevelt was too illustrious a man and so is the man we have in the White House tonight.

Dawes is Praised

Of the same caliber is General Charles Gates Dawes. After the war and the chaos in Europe, after five years of suffering and unrest following the Armistice did the people of that country go to Bob LaFollette to cure them of their ills? No! They went to Charles Dawes, the man who is going to be the next vice-president of the United States.

With such a country as this and such men as those willing to serve you as your president and vice-president respectively don't let anyone fool you about the wonders of Europe. Keep cool with Coolidge and return the republican party to power. Then you can sing with a happy voice, "My country tis of thee I sing."

Herbert Moore, boyhood friend and school mate of Calvin Coolidge told of the president's simple life in Vermont and of the influences that have guided him so well in the presidential chair. He sketched life in the little village of Plymouth, a village of only twenty-nine souls.

How He Took Oath

"When Coolidge learned of the sad death of President Warren Harding did he ask that the first justice of the United States Supreme Court come to Plymouth and administer to him the oath of office?" asked Moore. "No, he turned to his father and asked if he was still a justice of the peace." When answered in the affirmative the President said, "Father administer to me the oath of the United States" and it was done there in that little Plymouth, Vt farm house by the light of the kerosene lamp at 2:47 o'clock in the morning.

When he went to Washington did he ride on the palatial train that was being made up especially for him? No, he hitched his car to the regular train and went to his new duties without pomp or ceremony.

When his son Calvin, Jr. was working in a tobacco field down south and a fellow worker said: "Cal, if my father were President of this great country I wouldn't be working as a common laborer here in this tobacco field." What was his answer? It was "Comrade if your father were my father you would." That is the Coolidge spirit. No pomp or ceremony, no ostentation, just plain Americanism.

Apprise People

And so my friends, Moore concluded, We Plymouth friends of Calvin Coolidge have undertaken this little trip across the United States in order to apprise the people of the real facts concerning Calvin Coolidge – just fact's, that's all we want, for on facts the republican party will be returned to power by an overwhelming majority.

Hon. W. A. Jeffris, director of the tour and the man who placed General Dawes in nomination at Cleveland wound up the evening's program of speeches. He received the great accomplishments of the republican party in the field of foreign relations, its sweeping reductions in the national debt and equal slices in governmental expenditures and taxes, ridiculed LaFollette, spoke of the utter absence of any Davis tendencies encountered on the trip and wound up with several statements concerning the wealth and prosperity of the nation.

In speaking of the flag of the state of Iowa presented to the caravan at Clinton by the Daughters of the American Revolution, on behalf of the women of Iowa, Jeffries said: "That is a great motto you have on your state flag: 'Our liberties we prize and our rights we will maintain.' It is particularly appropriate in this campaign when those rights and those liberties are being attacked by such demagogues as the Senator from Wisconsin. If you believe in that motto you will see to it that Coolidge and Dawes are elected in November for in this campaign it's got to be either Coolidge or chaos."

Following the meeting a roster was taken of the Cedar Rapids men who voted for Abraham Lincoln. The list included: E. J. C. Bealer, John McAllister, J. S. Seymour, Milo P. Smith, C. F. Dunn, Eugene Alien, R. J. Wilson, George Bushnow and Sam Shaffer. These men in company with Congressman Cole and H. E. Spangler, chairman of the Linn county republican committee occupied the speaker's platform along with Atkinson, Moore and Jeffris.

Montrose Hotel, Cedar Rapids, Iowa

Waiting for lunch to be served by the Rotary Club at DeWitt. L. to R. - "Ab" Jefferis; L. L. Hasbrouck, J.P. Cowan-Manager; "Bill" Wallace; C. T. Doyle-Secretary; S. E. Mollden and E. C. Hoch

A third newspaper accounting reports the caravan event this way:

Parade Thru Streets Lines With Throngs. Half of 5,000 Mile Journey Across Continent is Completed With Stop Here. To Marshalltown Tuesday Night. All Day Drive From Sterling, Ill., Tiring One, But Spirits soon Revived On Arrival

Rolling into the city to the tune of "Oh, We're From Iowa," the Coolidge-Dawes Home-Town Club caravan struck Cedar Rapids in a blaze of glory at 7 p.m. yesterday and congratulated itself and the Cedar Rapids committee upon the excellent reception tendered it and upon the fact that it had completed exactly one-half of its 5,000 mile journey across the continent.

A trifle weary from the long journey they had made since leaving Sterling, Ill. Monday morning, eyes filled with dust flying up from the dirt roads they had encountered since leaving DeWitt, and throats parched and hoarse by the many speeches they had made during the day, the members of the caravan thrilled to the scene that greeted them at First Avenue and Twenty-seventy Street where fully 200 cars had gathered to meet them and escort them on a parade through the city.

Horns Blow

With horns blowing, pennants and banners flying and spirits revived, the caravan headed by the official welcoming party of H. E. Spangler, A. S. Reed and H. M. Wilson, American Legion bugler started on its journey down First avenue and through the business section of the town. Behind the caravan of six cars, including the truck on which was pictured the boyhood home of Calvin Coolidge, came 225 local motorists with cars packed to overflowing and horns and noise-makers doing overtime as though it was their specific duty to apprise the rest of the Cedar Rapids citizenary of the parade.

Down First avenue came the caravan. At every corner traffic was held up for a full five minutes while the giant parade filed by. Along the curbing were lined hundreds of other cars and out of the windows of the palatial and beautiful homes on upper Second avenue, leaned boys, girls, women and men, all waiving and shouting a welcome to the Vermont travelers. As the lights of the business district come into view the traffic became more and more congested and put the vanguard of Cedar Rapids motorcycle policemen to its best efforts to keep the passage clear.

Crowd Grows Thicker

At Tenth street the drivers shifted to intermediate to keep down the pace of their political carriages and as the caravan slowed down men, women and children ran along side the cars to gather the Coolidge-Dawes pennants that were dropped off at frequent intervals. Once across the railroad track the streets presented a united front of curious spectators and the drivers went into low. At the corner of First avenue part of the caravan lost the lead and headed south to Second avenue and over the narrower bridge. But it rejoined the leaders again and went honking and blaring down first street west trying to outdo the noise and pep of the Hanford post American Legion band. Down to Sixteenth avenue west it went, across the Sixteenth avenue bridge to Third street, and up Third street to Third avenue where it disbanded.

It was a glorious close to an even more glorious day.

Met at Lowden

Leaving Sterling, Ill. at 9 a.m. the caravan hustled along stopping only at Morrison, Clinton and De Witt before hitting the Cedar Rapids territory at Lowden.

At Clinton the caravan was presented with the Iowa flag and made a brief stop around the town square. At De Witt hungry stomachs and more thirsty mouths were appeased before the caravan preceded on to Lowden.

At Lowden began the series of frequent stops that were made between there and Cedar Rapids. After exchanging greetings with Sam G. Armstrong, Congressman Cyrenus Cole and a representative of the Cedar Rapids Republican, the leaders conducted a short program of speeches that was met with frequent volleys of applause. Hon. Jeffris, Herbert Moore and Congressman Cole were the speakers there.

Then the Coolidge-Dawes boosters were off to Clarence. As they sped along over well dragged roads they were greeted with a wave from farmers in the fields, a shout from their families on the homestead porch, and from little boys and girls returning home from school. Here and there a startled rabbit darted across the road and on a little further a scolding bluejay shouted a warning to cut down the speed.

At Clarence

Into Clarence sped the distributors of the Coolidge-Dawes gospel and here a larger crown estimated at 200 persons greeted them and listened with attentive ears to the same three speakers.

Nearing Stanwood the caravan was met by the official Cedar Rapids welcoming party. Three more speeches were delivered here and amid cries of "good-bye and good luck," the caravan headed out for Mechanicsville only to repeat the performance to an even larger crowd.

At Lisbon, the next stop, Major Dent Atkinson took the stump, in this case the running board of a Dodge automobile, and added to the other words of praise for Coolidge.

At Mount Vernon a crowd of close to 500 persons greeted the caravan and applauded enthusiastically when the speakers flayed LaFollette and extolled the republican nominees.

Buttoning up their coat collars and stamping their feet to warm them for the longer jump to Marion the members of the caravan piled into their cars and were off again. Just outside of Marion they were met by a big delegation of citizens of that city who escorted them into town where another and even larger meeting was held.

Then came the last lap of the day's journey and the reception at First avenue and Twenty-seventh street.

This morning the caravan headed out for Marshalltown, the next overnight stop on the tour. Tomorrow night the Vermonters will stop at Denison and Thursday they will cross the Missouri river into Nebraska and add another state to their conquests for Coolidge and Dawes.

Only eleven of them, but eleven good men and true, compose the caravan. They are: Hon. W. A. Jeffris, Herbert Moore, Major Dent Atkinson, John B. Farmer, W. A. Jeffris Jr, J. P. Cowan, Charles T. Doyle, William Wallace, S. E. Mollden, E. A. Sidell and E. C. Hoch.

L. L. Hasbrouck of Des Moines is accompanying the caravan across Iowa. Congressman Cyrenus Cole will accompany it on the remainder of the jaunt through the fifth congressional district and at Omaha Senator James Wadsworth of New York will join the party.

The *Cedar Rapids News* reported the next day:

Shows Human Side of Calvin to Local Folk. school Friend Of the President Talks to Mass Meeting In City.

An audience estimated at between 1,200 and 1,500 persons, that filled the City auditorium last night, heard of the advantages of safe and sane America, of the home life and the personal, human side of Calvin Coolidge, and of the accomplishments of the Coolidge-Dawes trans-continental caravan that reached the city at 7 p.m. yesterday.

The speakers – W. A. Jefferis of Omaha, the man who nominated Gen. Charles G. Dawes at the Cleveland convention; Herbert Moore, Plymouth, Vermont, who attended district school with Coolidge and Major Dent Atkinson, world traveler and lecturer, were heard with close attention and interrupted frequently by applause.

Major Atkinson, the first speaker, introduced by Representative Cole who presided, spoke briefly but forcefully of conditions as he has found them in devastated and bolshevistic Europe and of the marked contrast presented by good government, good working conditions, American liberties and ideals which folk enjoy in the United States.

Why Caravan Was Organized

Herbert Moore, the farmer of the tour, who left the fields early in September to join the little band of campaigners and tell the folks along the Lincoln highway of his friend, Calvin Coolidge, then told why the caravan had been organized.

"It was because Robert M. LaFollette, up there in Wisconsin, made the statement that Calvin Coolidge was a cold-blooded aristocrat and never saw a farm," exclaimed Moore. "The home town folk of Mr. Coolidge were indignant. We didn't like it. So here we are going across the continent for the purpose of acquainting the people with the exact facts of the boyhood of Mr. Coolidge."

"Mr. Coolidge was born on a farm and reared on a farm. Though he is not a farmer and has never been one, he understands the farmer's problems, he knows how much toil is necessary to make an honest dollar on the farm, and though the most elaborate summer resorts were offered him for his vacation – where did he spend it? It was back on his father's farm. Just where he had spent many other vacations."

Mr. Moore contrasted the responses of the three major candidates to their official notification of the nomination. Mr. Davis, he said, knocked the republican party and LaFollette knocked both the republicans and the democrats.

"But President Coolidge never mentioned the democratic party in his speech, he didn't refer to LaFollette, but he did get down to brass tacks and say what he would strive to do if elected."

As a striking indication of Mr. Coolidge's strict sense of attending to business and keeping his nose out of other's affairs, of his duty as he sees it, occurred at the time of the Boston police strike when Mr. Coolidge was governor of Massachusetts, Mr. Moore said. Though Coolidge's friends told him his action would mean the end of his career, he replied, "There can be no compromise between right and wrong."

Mr. Moore closed his talk with the poem that he uses so tellingly to conclude each speech and which is printed elsewhere in this issue.

Attacks LaFollette

Mr. Jefferis, the principal speaker who closed the rally, paid little or no attention to the democratic party. The share of his speech devoted to attack was launched against LaFollette and La Follettism. He pled for maintenance of the fundamentals in American life and government and scathingly denounced LaFollette's attack on the powers of the supreme court.

If he and his supporters should win in their attack against the Constitution, they would make it possible for congress to enact a law over and above the Constitution and the supreme court. It might enact a law whereby our religious liberties would be taken away; it might do away with the freedom of the press, the freedom of speech, the right of trial by jury, state's rights in their domestic affairs; it might enact a law abolishing habeas corpus and the rights of property – rights and liberties which men have fought and bled for.

"If we love our institutions and our liberties, we should think well before we follow such doctrines as these. La Follettism is antedeluvian, it is turning civilization back to chaos and to cave-man days of the autocracy of force," exclaimed Mr. Jefferis.

After reciting statistics indicating the magnitude of American products in the world's storehouse of food and machinery, Mr. Jefferis cited some of the accomplishments of the republican administration of Harding and Coolidge. The cost of government has been reduced by many millions, he said, under the economy of the budget plan. In discussion of the latter he praised the work of James W. Good, former representative from the Fifth district, who was the author of the measure.

Those Peace Treaties

Fifty treaties of peace have been made by the administration, he declared, forty-six of which have been ratified by the senate. Then there was the limitation of armaments conference and the Dawes reparations plan.

"Oh what a contrast," he said. "When the last democratic administration sent a commission to Europe to help settle the troubles over there, it compromised a whole boatload, a ship was chartered for that special purpose. But President Coolidge merely authorized three good business men who went over there and talked common sense to them."

Mr. Jefferis denied that the farmers are being taxed to help eastern industries under the Fordney protective tariff. Farm implements, fertilizer and many other things used by the farmer are on the free list, he said.

Concluding his speech with the state flag, presented the caravan at Clinton by the D.A.R. held before him, Mr. Jefferis read its motto, "Our liberties we prize and our rights we will maintain."

"I thought how appropriate that is in a campaign like this," he said. "The women who presented this flag, sensed a doctrine that is being preached against American ideals... Women of Iowa, men of Iowa, if you believe in the motto of your state, you will see that this contest ends with Coolidge and Dawes triumphant. If it doesn't, it means chaos with the election of the President thrown into the house of representatives and then the senate, with the best chances there for Charles Bryan and we don't want Bryan."

On the platform were eight Civil War veterans who voted for Abraham Lincoln – John McAlester, Judge Milo P. Smith, E. J. C. Bealer, George S. Bushnell, R. J. Wilson, Eugene Allen, C. F. Dunn and J. S. Seymour. Large portraits of Coolidge and Dawes on which lights were directed were placed at one corner of the platform.

Tuesday, September 30
Day 22

Left Cedar Rapids, Iowa at 9:00 a.m.

Log: Stops at Ames and Marshalltown, 85 miles. Spent night at hotel in Marshalltown.

Review: Congressman "Cy" Cole is an added speaker through Iowa. Big reception at Iowa State University in Ames.

Night meeting in Odeon Theatre.

—E. A. Seidel

National Coolidge College Clubs button shows support for the Coolidge campaign. 7/8"

Iowa State University band leads the Coolidge Caravan parade through Ames

The *Marshalltown Times Republican* for September 30 filed the following special report:

IOWA WELCOMES MOTOR CARAVAN

Coolidge-Dawes Boosters Greeted by Various Towns and Cities, Hundreds in Escort Into Cedar Rapids

Cedar Rapids – Iowa towns along the Lincoln Highway gave the Coolidge-Dawes caravan an enthusiastic welcome in its first day in the state starting at Clinton and reaching Cedar Rapids last night. This city is the half-way mark in both distance and time, since the caravan left Plymouth, Vt., President Coolidge's birthplace, Sept. 9.

At every town a crowd of from 150 to 300 gathered around the six cars in the caravan and listened eagerly to W. A. Jefferis, director of the tour, Herbert Moore, boyhood friend of President Coolidge, and Maj. Dent Atkinson, the speakers of the tour. Every town was decorated with flags and at Stanwood all the school children were assembled to greet the caravan.

At Marion twenty-five automobiles headed by Mayor A. E. Crewe escorted the Coolidge-Dawes cars into that town. From there to Cedar Rapids there was a constant stream of cars. Here 225 automobiles decorated with flags and banners and containing more than 800 persons formed a parade that brought the caravan into town between solid lines of parked automobiles, while the American Legion band and bugle corps at the head of the procession and hundreds of sirens provided the music.

Representative Cyrenus Cole met the caravan at Lowden and accompanied it today to Marshalltown. H. E. Spangler, A. S. Reed, a legion bugler, and newspaper men met it at Stanwood, thirty miles east of here, and acted as pilots into the city.

The parade about the city was followed by a mass meeting in the auditorium, where Representative Cole presided, introducing Mr. Jefferis, Maj. Atkinson and Mr. Moore. The latter made a distinct hit at every town as the only genuine farmer who ever stumped the country in a presidential campaign. His stories of the boyhood of the president pleased the people. He refuted LaFollette's charge that President Coolidge is a wealthy aristocrat, telling of the president's birthplace and early days on the farm. Though not a farmer, the president understands the farmers problem, Mr. Moore declared.

Mr. Jefferis took up the accomplishments of the republican administration, and told of the enthusiastic reception given the caravan all along the way, predicting a Coolidge victory.

Major Atkinson dealt with the conditions in Soviet Russia in comparison to the United States and told his hearers that if they valued continued good government and American rights and liberties, growing prosperity and the perpetuation of American ideals, they will vote for Coolidge and Dawes.

A. W. Jefferis gets ready to address audience from vehicle running board at street meeting in Louden, Iowa

Crowd at Caravan meeting at Iowa State University in Ames as Major Atkinson gives address

Wednesday, October 1
Day 23

Left Marshalltown, Iowa at 9:00 a.m.

Log: Stops at small towns across Iowa all the way to Dennison. 140 miles. Spent night at Hotel Dennison in Dennison.

Review: Mrs. A. W. Jefferis joined us here; also Ed Black of the "Omaha Bee."

Hotel here is terrible. I think the bed-bugs are holding their annual convention.

—E. A. Seidel

Edward Black, staff correspondent for *The Omaha Bee*, filed the following story from Denison on October 2 discussing Herb Moore's speech about Coolidge's earlier years:

Herbert L. Moore, neighbor of Calvin Coolidge and one of the speakers of the Coolidge-Dawes Lincoln caravan, attended school with the president at Black River academy.

Moore and the president occupied the same room while at the academy. "There was a library of 150 books in the house where we roomed," Moore said. "Calvin read every book and did not idle his time during the evening. We fished and hunted together. My father and I helped Calvin and his father on their farm and they helped us."

This is Moore's first trip west of New York state. "I observe that the girls are stouter out here than in New Engand," he said. He stated at breakfast this morning that when John Coolidge wishes to speak to his son at the White House over the telephone he must go to the grocery store. There is a reason. "Aurora Pierece, housekeeper for John Coolidge, will not answer a telephone, but she is one of the slickest housekeepers you ever saw. She is up at 4 every morning and at 5 the old gentleman is called to breakfast."

Moore told of the village of Plymouth in the township of Plymouth and the birthplace of Calvin Coolidge. "You go from Ludlow up the Black river valley 10 miles and turn to the right up a mountain that rises 200 feet in one-eighth of a mile and you come to Plymouth village, which has a population of 29 when I am home."

Newsprint glimpses of the Coolidge-Dawes Caravan tour activities: A group of girls holding some of the state flags presented to the Caravan; W. J. Jefferis, director of the tour, in a characteristic speaking pose; Jefferis and Herbert Moore, boyhood friend of Coolidge, sitting on log waiting to be called on for a speech; Old timers looking over the roster of men who cast their first vote when Lincoln was elected in 1860

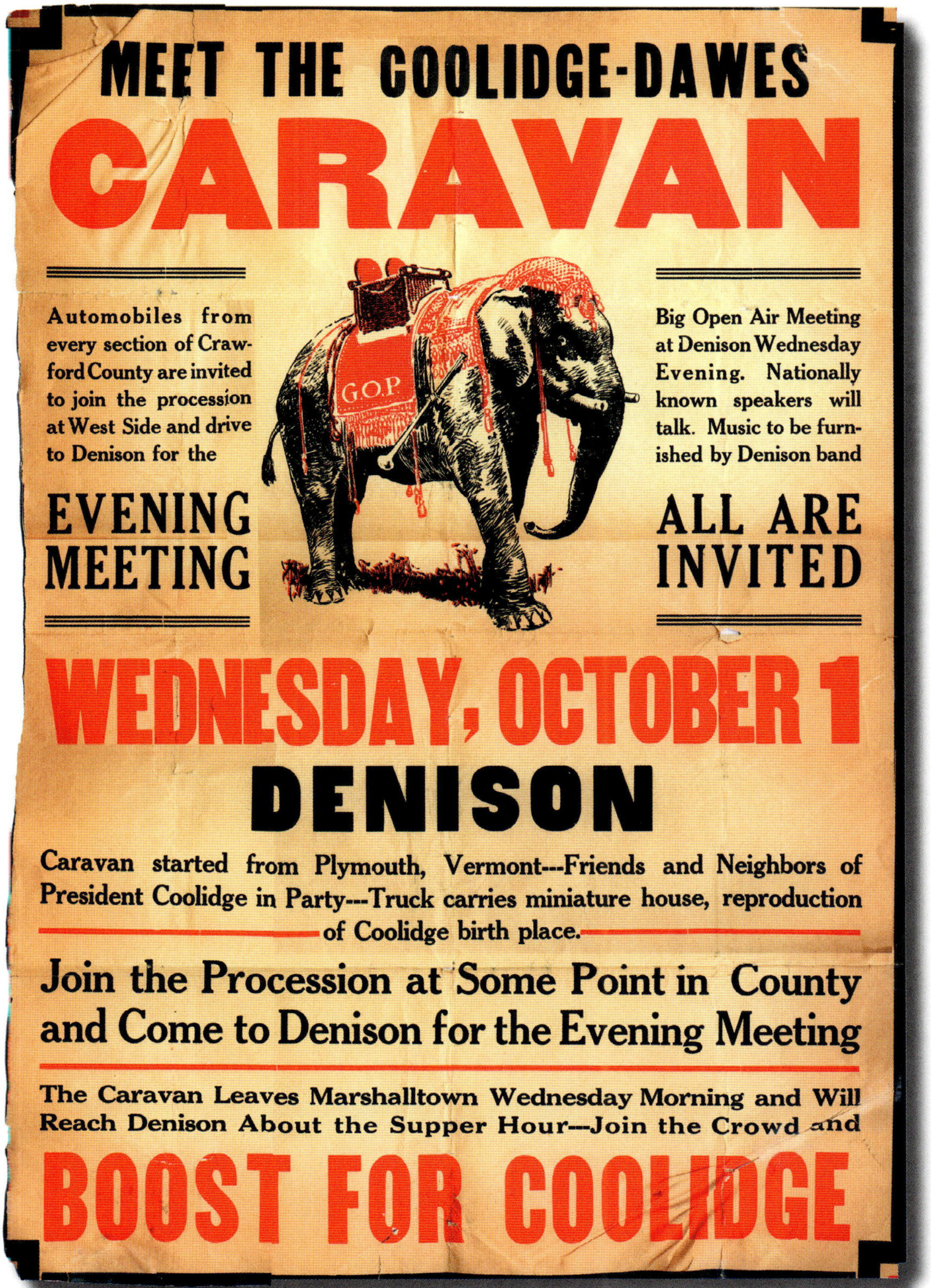

Large colorful posters draw crowds to big open air evening meeting in Denison

Apparently Ed Black, as a newspaperman, was rather "taken" by Herb Moore. He filed the following feature on Moore the following day upon the caravan's arrival in Omaha:

Coolidge-Dawes Caravan Centers Around Boyhood Friend of President. "Farmer" Herb Moore Privileged Character at White House; First Trip West of Gotham.

There are few persons who have the temerity to walk into the White House and address the president as "Calvin."

One is Col. John Coolidge, father of the president and another is "Farmer" Herb Moore, neighbor of the Coolidge family for more than half a century, and playmate and schoolmate of the chief executive of the nation.

Mr. Moore is the unique and outstanding figure of the Coolidge and Dawes caravan which arrived here Thursday afternoon, on its transcontinental tour from the Green Mountain state to the Golden Gate. Omaha is honored in having Albert W. Jefferis as director of the tour, but from the human interest standpoint Herb Moore holds first place. He knows as much about Calvin Coolidge as any other man in America.

Were Roommates

It happened that Moore and the president were born on adjoining farms and are about the same age. They attended the same district school and later went to the same academy, sharing the same room. Herb and his father helped Calvin and his father on the farm in times of need and the courtesy was returned in kind.

When Senator LaFollette recently stated that President Coolidge was a cold-blooded aristocrat and never lived on a farm, there was indignation in Plymouth, Vt., birthplace of the president.

Plymouth, Vt., comprises four farm houses, two small tenant houses, a cheese factory, one store, Congregational church and a school. There are 29 residents, including Herb Moore, who is census enumerator. An indignation meeting was held and Moore was named as chairman. The coast-to-coast automobile caravan was the result. These pilgrims were determined to tell the people of the life of Calvin Coolidge and they knew of no better man than Herb Moore to tell the story. And they looked to Omaha for a man to direct the tour.

First Trip West

This is the first journey of Herb Moore west of eastern New York. The great west is unfolding its panorama of wonders to this man of the Vermont hills. He is having the time of his life. "The most interesting scene I have observed along the way was a green patch of clover in Ohio, with a lot of white and black pigs having the time of their life," said Mr. Moore. "You know, in Vermont we farm on the hillsides. It is interesting to me to see the great sweeps of land in Iowa and Nebraska. I observe some fine orchards between Missouri Valley and Council Bluffs, but I could not say much for the road between these towns. It was the worst we have experienced since we left Plymouth."

"I have heard of the livestock and packing house industry of Omaha and I wish we had time to visit the stockyards; but we must be getting along. We are due in Sacramento on October 18 and I expect to return by rail over the Canadian route and be home in time to vote."

He was reared in a Spartan school of industry and honesty. He wants the facts. He stated yesterday that he fears he will talk too much and that some of the accounts will get back to Plymouth and that Aurora Pierce or Dick Brown may take him to task when he gets back home.

Simple Family Cemetery

Mr. Moore told about the Coolidge family cemetery plot, one-fourth of a mile south of the village of Plymouth. There are 24 graves of the Coolidge family in this resting place. Six generations are represented. The oldest in this eternal camping ground is Captain John Coolidge, great-great grandfather of the president. He settled in Vermont in 1759 and was in the revolutionary war. There is a great-grandfather, a grandfather, the president's mother, a sister and a son, making the six generations. The family has occupied the same farm for nearly 150 years.

"Farmer" Herb Moore is "just an old shoe," to use a New England expression. He is just folks. On Wednesday at Tama, Iowa, he met a man named Lyman Burt who was born in Plymouth, Vt., in 1829. This 95-year-old Iowan and Moore had a close-up chat. Moore knew Burt's brother, David, whose daughter married Eben Slack. Walter Slack, son of Eben Slack, and Moore were born on adjoining farms in Plymouth and there is only three days' difference in their ages.

"When we started on this tour," said Moore, "Jefferis and I did all of the speaking for seven days, and they kept us at it, day and night, in sunshine and in rain. We just had to get some help."

"Brought Up With Him"

"Yes, I was brought right up with him," was the ingenuous reply of Moore to a woman at Missouri Valley when she questioned him about the birthplace of President Coolidge. She said there are five votes in her home for Coolidge and Dawes.

Mr. Moore has been a member of the Vermont legislature since 1917. He has also been a member of the state fish and game department. During a week last summer he walked 92 miles in overseeing the game of the state. He is a farmer by training. He organized the Home Town Coolidge club which now has a membership of 15,000 throughout the United States.

"Keep Cool and Keep Coolidge" is the title of a poem Moore is reciting along the way, in his transcontinental tour. It is his first long ride on the choo-choo cars. It is a great treat for Herb Moore and he is doing the best he can to tell the people just what kind of a neighbor Calvin Coolidge has been. And Mr. Moore confided the information yesterday in Omaha that next March he will take the truck that was seen here yesterday with the pictures of the Coolidge farm home, and direct it down Pennsylvania avenue in Washington at the inauguration of Calvin Coolidge.

Housekeeper Forbids Telephone

Aurora Pierce is and has been for many years the faithful housekeeper at the farm home of John Coolidge. She will not permit a telephone in the house, wherefore the father of the president of the United States has to go to the grocery store if he wishes to speak to his son at the White House.

Who is Dick Brown? He owns the only automobile in Plymouth, home town of the president. Dick's brother, a traveling salesman, operated this car for about 100,000 miles and then gave it to his kin. Dick worked several years in an automobile institution in Detroit and knows a few things about auto mechanics. The car is 12 years old, but it runs.

But John Coolidge would not take a chance in Dick Brown's car. The president's father, according to Moore, has a brown horse with a star in its forehead. He himself hitches this horse to a closed buggy and drives 12 miles to Ludlow and return when he wishes to go to town. Last summer during the Cleveland convention well intentioned friends called at the Coolidge farm to ask the elder Coolidge to go to the big city to witness the nomination of his son. John Coolidge was out on the farm with this horse and said he was too busy harrowing to bother about going to a national convention.

Thursday, October 2
Day 24

Left Dennison, Iowa at 8:00 a.m.

Log: Stops at Dunlap, Woodbine, Missouri Valley, Council Bluffs and Omaha, Nebraska. 89 miles. Spent night at Hotel Fontanelle in Omaha.

Review: Flag presentation on the Missouri River bridge.

Big parade into Omaha, and through the main streets. Night meeting held in the Brandeis Theatre.

Senator Burton of Ohio joined the party here as an added speaker in place of Senator James Wadsworth, who was originally scheduled.

Mrs. A. W. Jefferis left party here to return to Washington.

Had first tire trouble of the trip at Council Bluffs.

Ak-sar-ben Carnival on this week. That is Nebraska spelled backwards.

—E. A. Seidel

Edward Black, a staff correspondent for *The Omaha Bee*, wrote the following feature on the caravans visit to Dunlap and Logan.

Iowans Enthused by G.O.P. Caravan. Auto Train Delayed One Hour at Dunlap by Crowds

Uproarious enthusiasm greeted the Coolidge-Dawes automobile carivan on its arrival here (Logan) at 11:30 a.m. today, one hour behind its schedule. The caravan will probably reach Council Bluffs at 3 p.m. today.

Crowds thronged the business part of town here, and the arrival of the carivan was the signal for cheers and prolonged blasts from automobile horns.

The carivan left Denison at 9 a.m., expecting to arrive here in less than two hours, but was delayed by the crowds which gathered at Dunlap and Woodbine.

Stops were made this morning at Denison, Dunlap, Woodbine and Logan. The stop at Woodbine had not been planned, but the caravan stopped to address the crowds which had gathered.

Women Give Encouragement

"Nothing has impressed so much on this trip as the interest and the activity which the women have demonstrated in their support of Coolidge and Dawes," ex-congressman Albert W. Jefferis, leader of the caravan, declared.

Jefferis referred to General Dawes as "one of the greatest Americans of all time," and declared that chaos and confusion would inevitably result if LaFollette and "his socialistic associates" should, by chance, succeed in their attempt to throw the election into congress.

Mrs. Jefferis joined her husband at Denison, and will accompany the party on its trip into Omaha this afternoon. Herbert F. Moore, addressing the gathering at Woodbine, declared that "LaFollette is running on six platforms at once."

Caravan Answer to "Bob"

"We people of Plymouth conceived this auto caravan when LaFollette referred to Coolidge as a full-blooded aristocrat who had scarcely seen a farm," Mr. Moore said. "We wanted the country to know the sort of a man Mr. Coolidge is."

Moore stressed the economy which Coolidge has introduced into national affairs, and at each stop made this morning he has recited a short poem, "Keep Cool and Keep Coolidge."

The car of Albert W. Jefferis carries the official flag of Iowa, which was presented to him as the expedition crossed the Mississippi river at Clinton. He will carry it until the caravan leaves Council bluffs, when he will receive the official Nebraska flag.

Congressman W. R. Green of Council Bluffs joined the caravan Wednesday, and introduced the speakers at each stop this morning.

The next stop of the caravan will be at Missouri Valley, where scores of Council Bluffs republicans will meet the expedition to escort it into Council Bluffs for its meeting at Bayliss park. this meeting will probably be at 3 p.m., after which the caravan will immediately proceed to Omaha."

"Welcome to Omaha" choice 1-1/4" Coolidge button

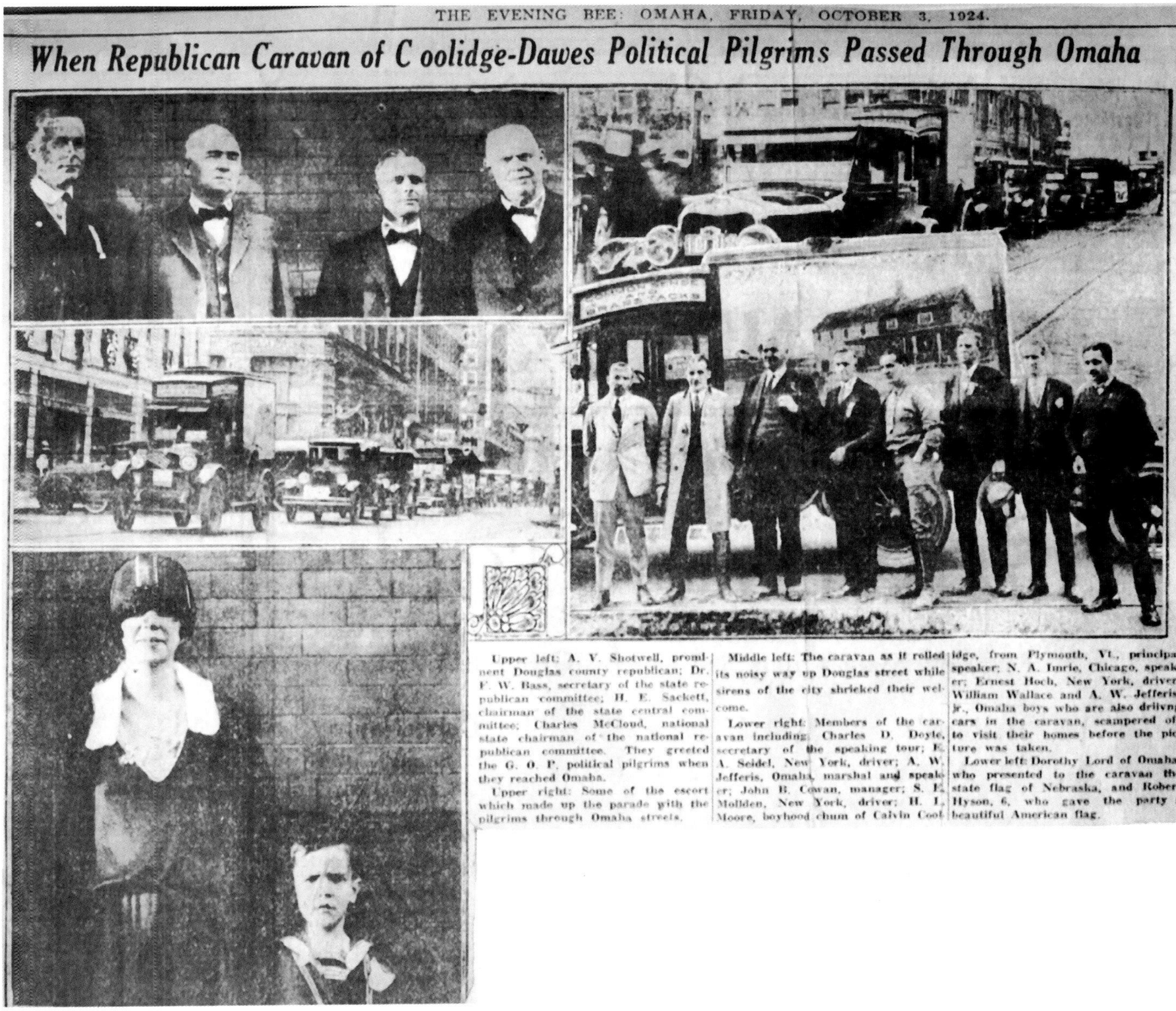

THE EVENING BEE: OMAHA, FRIDAY, OCTOBER 3, 1924.

When Republican Caravan of Coolidge-Dawes Political Pilgrims Passed Through Omaha

Upper left: A. V. Shotwell, prominent Douglas county republican; Dr. F. W. Bass, secretary of the state republican committee; H. E. Sackett, chairman of the state central committee; Charles McCloud, national state chairman of the national republican committee. They greeted the G. O. P. political pilgrims when they reached Omaha.

Upper right: Some of the escort which made up the parade with the pilgrims through Omaha streets.

Middle left: The caravan as it rolled its noisy way up Douglas street while sirens of the city shrieked their welcome.

Lower right: Members of the caravan including: Charles D. Doyle, secretary of the speaking tour; E. A. Seidel, New York, driver; A. W. Jefferis, Omaha, marshal and speaker; John B. Cowan, manager; S. E. Mollden, New York, driver; H. L. Moore, boyhood chum of Calvin Coolidge, from Plymouth, Vt., principal speaker; N. A. Imrie, Chicago, speaker; Ernest Hoch, New York, driver. William Wallace and A. W. Jefferis, Jr., Omaha boys who are also driivng cars in the caravan, scampered off to visit their homes before the picture was taken.

Lower left: Dorothy Lord of Omaha, who presented to the caravan the state flag of Nebraska, and Robert Hyson, 6, who gave the party a beautiful American flag.

Omaha greets the Caravan. Upper left: A. V. Shotwell, prominent Douglas county Republican, Dr. F. W. Bass, Secretary of State Republican Committee; H. E. Sackett, Chairman of the State Central Committee; Charles McCloud, National State Chairman of the National Republican Committee, greeting the G.O.P. political pilgrims when they reached Omaha; Upper right: Some of the escort which made up the parade with the pilgrims through Omaha streets; Middle left: The Caravan as it rolled its noisy way up Douglas street while sirens of the city shrieked their welcome; Lower right: Members of the Caravan including: Charles D. Doyle, secretary of the speaking tour; E. A. Seidel, New York driver; A. W. Jefferis, Omaha, marshall and speaker; John B. Cowan, Manager; S. E. Mollden, New York, driver; H. L. Moore, boyhood chum of Calvin Coolidge, from Plymouth, Vermont, principal speaker; N. A. Imrie, Chicago speaker; Ernest Hoch, New York, driver. William Wallace and A. W. Jefferis, Jr., Omaha boys who are also driving cars in the Caravan, scampered off to visit their homes before the picture was taken; Lower left: Dorothy Lord of Omaha, who presented to the Caravan the state flag of Nebraska, and Robert Hyson, 6, who gave the party a beautiful American flag

The Council Bluffs newspaper gave the caravan event a huge headline: **"G.O.P. CARAVAN ARRIVES AT BLUFFS. Triumphal Entry Made by Political Pilgrims Between Cheering Lines."**

Heralded by an advance guard of 50 escorting automobiles, the Coolidge-Dawes caravan arrived in Council Bluffs at 2:35 p.m. Thursday.

Factory whistles and motor horns welcomed the political pilgrims noisily from the moment they were first sighted until the parade entered the business district and stopped at Balliss park for an open-air mass meeting.

The caravan arrived in Council Bluffs almost a half hour ahead of the time it was expected, as it was an hour behind its schedule when it left Logan, Ia, and Council Bluffs was at Missouri Valley, where two speakers addressed an audience of about 500.

Triumphal Entry

The parade made a triumphal entry into Council Bluffs between rows of cheerers who lined the streets in the business section. Considerably more than 1,000 people gathered at Bayliss park for the mass meeting.

At the opening of the mass meeting, Congressman Albert W. Jefferis, leader of the caravan, asked those of his listeners who had voted for Lincoln to sign the Lincoln register following the meeting.

Charles E. Swanson, chairman of the Republican Central Committee of Pottawattamie county, then introduced Herbert C. Moore of Plymouth, Vt., who delivered the principal address of the meeting.

Delayed by Puncture

Following the speeches, the speakers prepared to leave at once for Omaha.

Several hundred Council Bluffs republicans accompanied the caravan into the city. Fifty cars, gaily decorated with Coolidge-Dawes slogans and banners joined the caravan at Missouri Valley and along the route into Council Bluffs.

Congressman Jefferis' car was delayed for a few minutes by a puncture half way between Council Bluffs and Missouri Valley. It was the first tire trouble which any of the seven cars in the Caravan has encountered since leaving Vermont on September 9.

Hotel Fontanelle, Omaha, Nebraska

The keynote speaker at the Republican National Convention in Cleveland, Representative Theodore E. Burton, arrived in Omaha earlier in the day, preparing for his role as main speaker at the caravan's rally in Omaha on Thursday night. The Omaha newspaper carried the following article:

Burton Sees Big Coolidge Vote. G.O.P. Keynoter Here Flays Brookhart for Attack on Dawes

Representative Theodore E. Burton of Ohio, keynote speaker at the republican national convention last June, took exception on his arrival in Omaha Thursday to Senator Brookhart's attack on General Dawes.

The Iowa senator, in a letter to the republican national committee, attacked the party's vice-presidential nominee for his utterances since the convention, and also attacked General Dawes because of his connection with the famous Lorimer bank failure in Chicago a few years ago.

Brookhart demanded that Dawes' resignation be asked and a farm bloc leader, preferably Senator Norris of Nebraska, be chosen as successor.

Brookhart Letter Unjust

Representative Burton characterized the Brookhart letter as "utterly unjust and an appeal to the unthinking." He explained that the transactions between Dawes and Lorimer's banks were according to law and known to the bank examiners at the time. Some years later, a careful appraisal showed that funds deposited by the Lorimer institution in Dawes' bank were $108,000 short of the estimated value of securities, he said. At no time, he declared, was General Dawes under a cloud.

The republican keynoter said he felt assured that Coolidge and Dawes would win by a large margin.

"The republican ticket will be victorious by a large majority in Ohio," he said. "Many persons believe it will exceed that of Harding four years ago. I have also visited Pennsylvania, where even the most optimistic democrats have given up hope."

Greeted by Nephew

Representative Burton, who will speak at the Brandeis theater tonight in connection with the Coolidge-Dawes caravan, was greeted at the Union station on his arrival at 11:35 by his nephew, William M. Burton, president of the Douglas County Coolidge-Dawes club.

The other members of the reception committee were John L. Kennedy, president of the United States National Bank, who served with Mr. Burton in Congress, National Committeeman Charles A. McCloud of York, Joseph W. Marrow, organizer of Coolidge-Dawes clubs; Representative W. G. Sears, his secretary, Jack Lee, and Harry S. Byrne, member of the republican state committee.

Burton was taken to Hotel Fontenelle and will be guest of honor at a dinner tonight. After his campaign speeches in the west he will invade the border states of Kentucky and Tennessee and also devote some time in Michigan and Indiana. He intends to spend the last few days of the campaign in Ohio.

Representative Burton recalled that his several visits to Omaha in recent years have been pleasant. He especially remembered his reception here nine years ago when he was being prominently mentioned for the republican presidential nomination in 1916.

The Omaha Bee carried the following article the day the caravan arrived in the city:

G.O.P. Caravan To Get Royal Welcome Here. Motor Train Due in Omaha at 5:30 p.m.; Dinner at Fontenelle; Rally at Brandeis. Burton Arrives for Speech.

Between 500 and 1,000 automobiles owned by republicans are scheduled to meet the Coolidge-Dawes motor caravan at the west end of the Douglas street bridge at 5:30 Thursday afternoon and escort the visitors through the principal business streets.

Many republicans motored to Council Bluffs and Missouri Valley, Ia., to meet the caravan.

Congressman Theodore E. Burton of Ohio, who delivered the keynote speech at the republican national convention in Cleveland last June, arrived at 11:35 a.m. over the Chicago & Northwestern. He was met by a delegation of Nebraska republicans headed by his nephew, William M. Burton, an Omaha lawyer.

Bluffs Delegation Large

A delegation of nearly 500 republicans from Pottawattamie county will motor to Missouri Valley and Logan to escort the transport train into Council Bluffs. The caravan is scheduled to arrive there at 3 p.m.

Whistles are to herald the approach of the caravan when it enters Council Bluffs. Upon its arrival the auto train will circle Bayliss park and stop for a half hour on the Sixth street side of the park, while addresses are being delivered by party leaders.

Members of the caravan, which is on route from President Coolidge's home town of Plymouth, Vt., to San Francisco over the Lincoln highway, and Representative Burton will be guests at 6 tonight at a dinner at Hotel Fontenelle. It is expected that about 150 Omaha republicans will attend the banquet, which will be held in the palm room.

Meeting at 8 p.m.

The cars of the unique political caravan will be parked in front of the Brandeis theatre, where Burton will deliver an address tonight at 8. The caravan will resume its westward journey Friday, going to Columbus and on to Kearney Saturday and will rest there Sunday. Monday it will go to North Plette and Tuesday to Sidney.

Friday, October 3
Day 25

Left Omaha, Nebraska at 10:00 a.m.

Log: Stops at Valley, Freemont, Schuyler and Columbus. 90 miles. Spent night at Hotel Evans in Columbus, Nebraska.

Review: Had another flat 10 miles out of Omaha.

We were the guests of the Freemont Rotary Club for luncheon.

Left caravan at Columbus to spend week-end at Norfolk, Neb., Hoch driving Dodge and Wallace driving Reo.

—E. A. Seidel

The Omaha newspapers offered this feature on the caravan's visit to their city:

G.O.P. Auto Train Leaves for West. Coolidge-Dawes Tourists Well Pleased With Reception in Omaha

The Coolidge and Dawes Lincoln highway automobile caravan pushed out into the west Friday morning at 10, to cross Nebraska, the 11th state of the trip, which was started September 9 at Plymouth, Vt.

After a night's rest at Hotel Fontenelle the members of the party were quite "peppy," as expressed by John Y. Doyle, secretary of the organization.

O. S. Spiltman, attorney general of Nebraska, joined the party here and is scheduled for a series of talks during the Nebraska itinerary.

Albert W. Jefferis of Omaha, director of the tour, started with the party at the Coolidge farm home and will go through to the Pacific coast.

Pleased With Receptions

"We were pleased with the reception in Omaha," Jefferis said. "In view of the Ak-Sar-Ben festivities I think the demonstration was indicative of republican activity and of the interest being taken in the Coolidge-Dawes campaign. Through Iowa we were received by crowds everywhere. We had to stop at many towns which were not on the schedule. The entire trip through 10 states and along 2,800 miles of highways has been a great success."

H. L. Moore, known as "Herb" to the members of his party, expressed gratification over the meeting Thursday night in Brandeis theater.

Moore's Chauffeur Omahan

"Although this is the first time I have been west of New York State," Moore said, "you would be surprised to know how much I have learned about Omaha on my trip. My driver is William Wallace of Omaha, and all along the route he has been telling me about the city. I have read and have been told much of this great western empire, but one has to see this country of ours to appreciate its greatness. It just makes me believe that Columbus could not have missed it. When I get back to Plymouth my wife will say, I told you so because she came from North Dakota. It is my intention of returning home by the Canadian route on the railroad."

Albert W. Jefferis, Jr., remains with the party as one of the motor car operators.

At Columbus Tonight

The party will rest tonight at Columbus, is due to arrive in Kearney Saturday night and rest there on Sunday, will finish Monday's run in North Platte and complete the Nebraska leg of the journey Tuesday night at Sydney.

Postmaster Charles E. Black gave members of the party his personal attention Friday morning, the location of several bundles of laundry being of chief concern.

The first scheduled speaking stop of the party from Omaha was Fremont at 11:30 a.m.

Caravan stops to address group in Freemont, Nebraska

Herb Moore getting some "Big Ears" from a real Nebraskan

Congressman Simmons and Herb Moore pose at back of REO Speed Wagon

Another article on the Omaha visit:

Coolidge Friend Wins Audience. Farmer and Statesman Occupy Leading Roles in G.O.P. Rally

A quaint hamlet, nestling like a pastoral poem among the Green mountains of Vermont, left its mark of rugged honesty, loyalty and stability upon Omaha Thursday night.

A farmer from those hills spoke. He had spent all his life forcing a stubborn, stony loam to produce grain and hay and vegetables. Until a few days ago he had never appeared on a public platform, yet he addressed a large audience from the stage of the Brandeis theater and won his auditors by an eloquent simplicity. He has a message to deliver to the American people and he is tackling this task with the same determination that he has long used in dealing with less complex problems of the soil.

On the same platform was one of the most distinguished republican orators of the generation, former Senator Theodore E. Burton of Ohio, some years ago a candidate for the presidential nomination, and the keynoter of the last national convention. The Vermont farmer, boyhood friend of the president of the United States, gave color to the rally because of his sincerity and single-mindedness of purpose. The Ohio statesman treated his subject with the polished phrases and constructive logic that are his by virtue of years of experience. They both hit home.

Picturesque Figure

Herbert L. Moore of Plymouth, Vt., is the most picturesque figure of the Coolidge-Dawes caravan which is going over the Lincoln highway from the Green Mountains to the Golden Gate. This little group of Coolidge boosters from his old home town reached Omaha Thursday afternoon. The Vermonters attended the meeting at the Brandeis in a body.

Moore is about the same age as President Coolidge and they played together as boys. They were roommates at the old Ludlow Academy at Ludlow, Vt., and after that, the farmer explained last night, their trails parted. Coolidge went on to college and the practice of law. Moore returned to the farm. And now Moore, whose friendship for Calvin Coolidge has not wavered with the years, is doing missionary work all the way across the continent in one of the most unique and effective political maneuvers in history.

He explained that the home folks at Plymouth had decided to do what they could in telling the country why Coolidge should be re-elected. He told about the boyhood of the president, of the many qualities which eminently fit him for the office he now holds.

Coolidge or Chaos

"The people should be the ones to choose the president," he declared, "and not the congress of the United States."

Former Senator Burton said that it would be either Coolidge or chaos. He recalled the economic distress of 1876-77, during the Hayes-Tilden controversy, and appealed to the democrats to vote for Coolidge. The speaker declared that Davis did not have any show at all to win and that should the election be thrown into congress "it would be a calamity; the wheels of industry would stop."

Senator Burton made a strong plea for co-operative marketing and said that the mounting cost of living in recent years had been due to the widening gap between the producer and the consumer. He also asserted his belief that there would be a "back to the farm" movement within a decade, due to the great law of supply and demand and not because of any agitation.

Senator Burton Speaks

Taking up the Teapot Dome question, Burton said that the opponents of the republican party had sought to stir up an atmosphere of scandal and had ignored the real issues of the campaign. He condemned the action of Senator Fall in accepting a large loan while in office. The least, he said, may be found to have been illegal. He quoted figures, however, to prove that the leasing of government oil preserves had been a detriment.

"No one can prove," he said, "that the Teapot Dome lease was not a favorable bargain for the United States government." It resulted in building a $15,000,000 pipe line. The oil field has turned out to be only about one-tenth what it was thought to be worth, before drilling began.

The Coolidge-Dawes caravan, which has already traveled 10 states, is preparing a book containing signatures of men who voted for Lincoln. This book will be presented to President Coolidge. An interesting feature of the rally was the signing of this book by seven men who had cast in 1860 or 1864 for Lincoln. They were J. H. Berger, Fred H. Barber, Robert J. Wass, Dr. J. B. Ralph, F. A. Gallup and C. E. Welch, all of Omaha, and C. G. Farwell of Watana, Wis.

An American flag was presented to the caravan by little Bobby Hyson, son of Mr. and Mrs. P. C. Hyson. The flag of Nebraska was presented by Miss Dorothy Lord, and that of North Dakota by Miss Marjorie Morehouse. Former Representative A. W. Jefferis of Omaha who is director of the caravan, accepted the standards, and described the trip thus far. He will accompany the party to the Pacific coast.

O. E. Engler, chairman of the Douglas county republican central committee, presided. On the platform were numerous members of the state committee, including chairman H. E. Sackett, and other republican leaders, both men and women.

The Coolidge-Dawes caravan departs at 8:30 a.m. for other points west. It will be in Columbus tonight.

The Columbus report in *The Omaha Bee*...

Columbus Autos To Meet Caravan.

A delegation of 15 autoloads of Columbus men, representing the Platte county republican central committee, will leave Columbus at 2 p.m. Friday for Schuyler to act as escort to the Coolidge-Dawes caravan. At 8 p.m. here a public program in the downtown square will be given, to be preceded by a band concert.

United States Senator Wadsworth, New York; former Congressman Jefferis, Omaha, and E. C. Huston, Tekamah, Neb., republican candidate for congress in the Third district, will be among the speakers. The meeting will be followed by a torchlight parade.

Another Nebraska newspaper carries the following, interesting general article:

COOLIDGE CARAVAN GETS BIG WELCOME.

Greeted by Crowds All Along Route

The Coolidge-Dawes Lincoln automobile tour has assumed the proportions of a veritable triumphal march. At no time in American political history, according to reports from towns and communities already behind the tourists has such a warm and impartial welcome been given representatives of a political party as is now being given members of the republican caravan.

Welcoming committees and innumerable brass bands are vying with each other in an effort to provide the greatest din in the enthusiasm of the caravan's arrival at each state boundary. Towns along the route have put on their gala attire for the occasion. In the demonstrations, huge pictures of President Coolidge and his running mate, General Dawes, are being exhibited, not only from automobiles in the procession, but from every available building or bill board along the route of the Lincoln National highway.

Leading republicans of every state thus far visited have been unanimous in their assertion that the enthusiasm and acclaim which has been given the caravan stamps this original departure in electioneering as one of the most unique features ever attempted in a political campaign.

Saturday, October 4
Day 26

Left Columbus, Nebraska at 9:00 a.m.

Log: Stops at Silver Creek, Central City, Grand Island, Shelton, Gibbon and Kearney, Nebraska. 104 miles. Spent night at hotel in Kearney.

Review: Party joined by Congressman Simmons as an added speaker.

Dinner served at Methodist Church.

Kearney is the Hub of the Nation, 2,264 miles from Boston and 2,264 miles from San Francisco. "Herb" Moore presented with the hub and spokes.

Rejoined caravan here Monday morning.

—E. A. Seidel

Coolidge-Dawes Use Common Sense die-cut cardboard display (9-1/2" long)

Lead vehicle with driver, E. A. Seidel

Sunday, October 5
Day 27

Lay over at Kearney, Nebraska

Monday, October 6
Day 28

Left Kearney, Nebraska at 9:45 a.m.

Log: Stops at Overton, Lexington, Cozad, Gothenburg and North Platte, Nebraska. 105 miles. Spent night at Hotel McCabe, North Platte.

Review: Big meeting at State Normal School.
Dirt roads all through here.
Bill Pitton of Denver branch now driving Lincoln.
Every restaurant in town is run by Japanese.
—E. A. Seidel

Tuesday, October 7
Day 29

Left North Platte, Nebraska at 8:00 a.m.

Log: Stops at Hershey, Sutherland, Ogallala, Big Springs, Paxton, Chappel and Sidney, NE. 134 miles. Spent night at Hotel Park in Sidney.

Review: Weather is getting cold, and roads getting worse. Duck dinner by the Committee at the Union Pacific Hotel.
—E. A. Seidel

The *Keith County News* of Ogallala reported the following under the heading "**Caravan Speakers Addressed Big Crowd at Ogallala:**"

The Coolidge-Dawes Lincoln caravan arrived in Ogallala at noon Tuesday. Four speakers addressed the crowds assembled and the caravan continued west on the Lincoln Highway with Freeman Whitney, candidate for representative of the 94th district and C. W. Johnson, of Potter, candidate for state senator of the 31st district, piloting the cars to Big Springs.

Congressman Robert G. Simmons was with the caravan and made the announcement that he would be in Ogallala on Tuesday, October 28th.

Mr. Moors, of Plymouth, Vermont spoke intimately of Calvin Coolidge whom he has known since childhood, and asked the voters to show at the polls the confidence in Coolidge that he deserved.

Senator Pitts, of Iowa, told of the important things that the Republican party had done for the people, that it was a constructive party and always gave the country a sound business administration.

Ex-Congressman Jeffries, of Omaha, who is in charge of the caravan, told of the economies brought about by the Republican administration with the help of the budget system and the lowering of taxation.

Mr. Jeffries proved that the third party could not elect a president and warned the people that they would only dupe themselves if they failed to elect their president instead of causing the election to be thrown into the House of Representatives and the Senate of the United States.

We regret that we have not the space or time to give the congressman's address in full for many who heard him came away with words of praise for the able manner in which he presented the facts.

Congressman Jefferis at Potter, Nebraska stop with veterans who voted for Abraham Lincoln

Congressman Jefferis with veterans who voted for Abe Lincoln at Potter, Nebraska

"Keep Cool and Keep Coolidge" is the message as Herb Moore does his bit at the Potter, Nebraska gathering

"Just a Flat" somewhere in Nebraska - Senator Pitt and Charlie Doyle help Hock change the tire...by looking on

The Sidney Telegraph, toted as being Western Nebraska's oldest and biggest newspaper, and independent in politics, gave the caravan's visit less than passing marks in their October 10 issue in a front page story:

Eulogize Coolidge; Denounce LaFollette, Republican Caravan Evidently Fears Third Party More Than the Democrats

The Coolidge caravan has come and gone, but it is very doubtful if any or many new votes were made as a result of the visit. For those who expected to hear the issues of the republican party expounded, the meeting was indeed a huge disappointment. In fact the entire meeting was devoted to eulogizing the president and lambasting LaFollette. Only one speaker made reference to the democratic party, which would indicate that the republicans are very much scared of the third party. In fact former Congressman A. W. Jefferies went so far as to say that anyone who voted for LaFollette was un-American.

The caravan, consisting of about a dozen cars and a score of speakers, was met at Chappell by a half dozen Sidney cars filled with prominent republicans. The caravan was escorted to Sidney with much pomp, arriving here shortly before six o'clock.

At 6:30 a duck and prairie chicken dinner was served at the Union Pacific hotel, attended by the caravan party and between fifty and sixty Sidney citizens. M. W. Dimery was chairman of the dinner party. Brief talks were given by the speakers who repeated and enlarged at the meeting held in the district court room immediately following the banquet. the court room was filled with men and a few women. County Attorney R. P. Kepler presided at this meeting.

The first speaker was H. L. Moore, a boyhood friend of the president. Mr. Moore's talk was devoted entirely to telling the audience what a fine man the president is. He said that the president owned only a small piece of real estate, rented a house at $30 a month and that his character was unassailable, which was known and admitted.

The next speaker was Congressman R. G. Simmons, who also devoted his talk to telling wiat a fine man the president is. Bob got real dramatic and almost brought tears to his tender hearted listeners in his eulogy. He stated he would be back Friday night of this week and tell the people what's what in politics.

Senator Pitt of Iowa was the next speaker. Why he was placed on the program at all was more than most could figure out. He only took up time. Yet he was the only speaker who alluded to the democratic party, and then in a historical manner. At least his large bulk helps as a shock absorber for the cars as they speed along. He waxed loud and warm in a resume of political strife from time immemorial, which is a dead issue.

But the party held their best card for the last talk. He was former Congressman Jefferies of Omaha. Mr. Jefferies is an able talker and has been known to make some very able addresses, but his speech Tuesday night lacked logic or the president's so-called common sense. He, too, devoted his entire address to mud slinging at LaFollette. Many of the audience tired of the tirade against political enemies and left before the meeting was half over. In fact the entire program was spent in advertising the third party and easily showed that they are disregarding the democrats entirely as an issue and are attempting to scare the people into voting for Coolidge by telling what a mean man LaFollette is. There was very little constructive work accomplished for the republican party. None of the vital G.O. P. issues were discussed. They seemed to infer that the people should vote for Coolidge solely because of his alleged personal character.

Wednesday, October 8
Day 30

Left Sidney, Nebraska at 8:00 a.m.

Log: Stops at Pine Bluffs, Kimball, Lodge Pole and Cheyenne, WY. 130 miles. Spent night at Hotel Plains in Cheyenne.

Review: Roads very bad. Mud and rain and terrific headwinds make driving dangerous.

Mollden driving Dodge stuck in ditch for two hours.

Hoch was blown around on the prairie like a piece of paper.

Main road impassable so we had to cross cattle range for 40 miles, with a cowboy escort.

—E. A. Seidel

Although not on the official schedule, *The Potter Review*, Potter, Nebraska, October 10 edition carries a first page article: **"Coolidge-Dawes Caravan in Potter Wednesday."**

The much heralded Coolidge-Dawes caravan arrived in Potter Wednesday forenoon about 9:30 and stayed an hour. The caravan proper is composed of six cars which have traveled from Plymouth, Vermont, the birthplace of President Coolidge, and will continue on to the Pacific coast.

The caravan was accompanied from Sidney to Potter by a number of Potter and Sidney cars and was met here by a delegation of serval cars from Kimball.

The first car of the caravan was a Reo speed-wagon with the truck part bearing large paintings of the humble birth-place of Calvin Coolidge.

The speakers were introduced in turn to the large crowd assembled by C. W. Johnson, republican candidate for state senator. The first speaker was President Coolidge's boyhood chum. His plea was to counteract the statements of candidate LaFollette that Coolidge was an aristocrat. Robert G. Simmons, congressman, was then introuduced but did not give a long talk due to the fact that he will speak here today (Friday), between 3:00 and 4:30 p.m. A. W. Jefferis of Omaha was then introduced and he gave a fine talk, urging the people to vote for Coolidge and Dawes and to save the country from chaos. He stated that Davis and LaFollette could not hope to win but if the latter got enough votes a president would not be elected and that the choice of a president would then be given to congress.

The school children attended the speaking and gave several lusty cheers for the caravan and the speakers as they appeared.

Several accompanied the caravan to Dix, Kimball and Bushnell.

Cheyenne's Plains Hotel (on right)

Vermont Caravan is greeted by house made of corn in the middle of the street in downtown Sidney, Nebraska

A typical small town audience listens to the Caravan speakers in front of a Cafe in Big Springs, Nebraska

Thursday, October 9
Day 31

Left Cheyenne, Wyoming at 9:00 a.m.

Log: Stops at Laramie, Parco and Rawlins. 130 miles. Spent night at hotel in Rawlins.

Review: Lost splash-pan crossing Divide and motor went dead due to the wetting it got.

Towed into Laramie by Reo dealer, where pan was replaced.

Got two flat tires between Laramie and Parco.

Blew front tire while doing 57 miles an hour, but car managed to stay on the road.

14 hours on the road today. Crossed Continental Divide, elevation 8,835 feet, in 2 feet of snow!

—E. A. Seidel

Putting snow in radiator after hard pull in Wyoming

The Cheyenne newspaper carried the following account on the previous day:

Republican Tour is Undaunted by Muddy Highways. Caravan of Cheyenne Republicans Escort Tour

Despite rain and the muddy condition of the highway, a number of automobiles carrying prominent Republicans left the Chamber of Commerce building at 3 o'clock Wednesday afternoon and proceeded eastward to meet the Coolidge-Dawes Lincoln Tour caravan, which is due in Cheyenne late in the afternoon. Earlier in the day a Cheyenne party in eight cars left for Pine bluffs, to welcome the caravan at the Wyoming boundary and escort it to Cheyenne.

Beginning at 8 o'clock Wednesday evening, at Cathedral hall, speakers with the Coolidge-Dawes caravan will deliver addresses at a public meeting to which all are invited. Governor Frank E. Lucas will preside at the meeting. The speakers will be Harold L. Moore, boyhood friend of President Coolidge; Hon. A. W. Jefferies, who nominated Mr. Dawes at the Cleveland convention, and Hon. W. W. Pitt, of Iowa. Boy Scouts will act as ushers at Cathedral hall.

Committees in charge of arrangements for the reception of the Coolidge-Dawes caravan here and for the meeting at Cathedral hall follow:

Car committee: J. H. Walton, chairman; F. D. Dildine, Arthur Piggott, J. C. Gans and T. H. Baldwin.

Program committee: W. C. Bond, chairman, C. L. Titus, A. L. Leslie and William Dubois.

The caravan will leave Thursday morning for the West. It is enroute across the continent, via the Lincoln highway, in the interests of the presidential candidacy of Calvin Coolidge and the vice presidential candidacy of Charles G. Dawes.

Crossing the Continental Divide between Cheyenne and Laramie. J. P. Cowan poses by sign

"A Big Six" and "Little Jeff" pose at the stop at the Continental Divide

Caravan undaunted by Wyoming snows

Friday, October 10
Day 32

Left Rawlins, Wyoming at 11:55 a.m.

Log: Stops at Wamsutter and Rock Springs. 114 miles. Spent night at Park Hotel in Rock Springs.

Review: Bill Wallace drove the Reo today as I was all in.
Roads in very bad shape.
Picked up some poor devil on the road who hadn't eaten for two days and was lost in the snow.
—E. A. Seidel

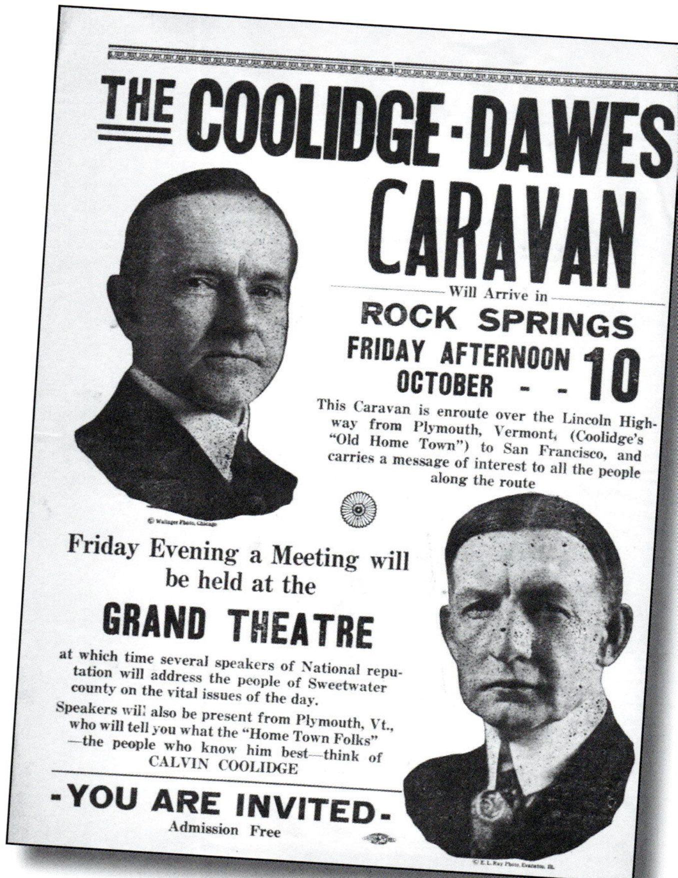

Local planning committees often announced the arrival of the Caravan and events through posters and fliers

Grand Theatre, Rock Springs, Wyoming (Courtesy of Rock Springs Historical Museum)

Park Hotel, Rock Springs, Wyoming (Courtesy of Rock Springs Historical Museum)

"Main Street" in Rawlins, Wyoming greets the Caravan from the east

Saturday, October 11
Day 33

Left Rock Springs at 9:00 a.m.

Log: Stops at Green River, Lyman, Fort Bridger and Evanston. 114 miles. Spent night at Evanston Hotel in Evanston.

Review: Driving very dangerous, average about 10 miles an hour.

Met Tom Lynch, an old schoolmate, dressed as a cowboy at Fort Bridger. Running a ranch at Lone Tree, Wyoming.

C. L. Welti of Salt Lake City, and Bill Pitton of Denver, both driving Lincolns.

Came through the Grand Canyon of Utah.

—E. A. Seidel

The Caravan easterners meet some cowboys in Fort Bridger, Wyoming

Taking a break in Evanston, Wyoming

Snow's over - taking off the chains in Green River, Wyoming

Front Street, Evanston, Wyoming

Senator Pitt views monument to General Bridger in Fort Bridger, Wyoming

Sunday, October 12
Day 34

Left Evanston, Wyoming (no specified time)

Log: Drove to Salt Lake City, Utah. 96 miles. Spent night at Hotel Utah, Salt Lake City.

"Trail Breakers of '24" - Caravan meets up with Official Lincoln Highway Trail Marking Car in Evanston, Wyoming

"Trail Breakers of '24" - Caravan meets up with Official Lincoln Highway Trail Marking Car in Evanston, Wyoming

Hotel Utah, Salt Lake City

Monday, October 13
Day 35

Left Salt Lake City, Utah (no specified time)

Log: Drove from Salt Lake City to Ogden and returned. 74 miles. Spent night at Hotel Utah, Salt Lake City.

The Salt Lake City newspaper reported the following events relating to the caravan's visit:

G.O.P. Boosters to Hold Big Rally at 8 o'Clock Tonight at Hotel Utah

Unofficially and informally, the Coolidge-Dawes transcontinental Lincoln tour is in Salt Lake City. Officially it will enter the state of Utah tomorrow, and will be welcomed at Ogden, where there will be a presentation of a state flag of Utah and speechmaking at noon. The afternoon will be spent largely in Davis county, and the caravan will reach Salt Lake at 6 o'clock in the evening.

Tomorrow the caravan will start westward, perhaps taking the Lincoln highway to Ely, Nev., although efforts are being made to persuade its members to follow the Zion Park highway, at least as far as Beaver, and then to swing west through Milford and Garrison to Ely.

Salt Lakers are invited, and Republican boosters are especially urged, to join in the automobile parade to Ogden, assembling on Third East street near South Temple street at 9 o'clock this morning. Captain J. Ray Ward, United States marshal, is taking charge of this part of the parade, and will distribute banners and distinguishing marks to the Utahns who join in the parade.

The caravan is due to reach Twenty-sixth street and Washington avenue, Ogden, at 11 o'clock and a ceremony of presentation of the state flag will occur there. There also will be assembled and specifically honored any persons who voted for President Lincoln who are present there. There will be speaking at the Egyptian theater, and a luncheon at Ogden, and the caravan will then start on its official tour of Utah, through Davis county, reaching First West and South Temple streets, Salt Lake, at 6 o'clock in the evening. The itinerary through Davis county already has been announced.

There will be a parade through Salt Lake streets, all who desire joining in, after meeting the caravan at First West and South Temple streets. Persons who were unable to devote the holiday to the trip to Ogden are especially requested to join in the evening procession. The route of the parade will be down Main street from the Brigham Young monument to Fourth South street, then to State street and north on State street to the Eagle gate, where the parade will disband. Several special features have been planned, including an industrial portion of the exhibit, featuring sugar, lead and wool.

Evening Meeting

In the evening there will be a big rally at the Hotel Utah, beginning at 8 o'clock. The speakers will include Congressman A. W. Jefferies of Omaha, who nominated General C. G. Dawes for vice president, who later notified him of the Cleveland convention's action and who has been with the tour from its start at Plymouth, Vt., the birthplace of President Coolidge. Other visiting speakers will be H. L. Moore of Plymouth, a boyhood playmate of the then future president of the United States, and M. W. Pitt of Crescent, Iowa. Mr. Moore is a farmer by occupation, and, without pretense at oratory, tells of incidents in the president's boyhood career and his life before he became president. Mr. Pitt was formerly speaker of the Iowa house of representatives.

The caravan left Plymouth September 9 and has now traveled close to 4,000 miles. It went first to New York, through Vermont, New Hampshire, Massachusetts and New York state, traveling down the Hudson valley, a trip of about 900 miles. Then it joined the Lincoln highway, and it has been on that thoroughfare ever since, with the exception of a few times when the members were persuaded to take a more circuitous route across a state for political campaign reasons.

"We have not been delayed an hour by bad roads," said John R. Cowan of Pittsburgh, Pa., manager of the tour, yesterday. "We have had some bad weather at times, too, and from Cheyenne to Laramie, Wyo., we traveled through a foot of snow. But, as a matter of fact, the worst trouble we had was on the first day out of Plymouth, when it was raining a regular cloudburst."

230 Meetings

"We have held something like 230 meetings since September 9, and have addressed more than 175,000 persons. In Chester and Lancaster counties, in eastern Pennsylvania, the procession which accompanied us was fifteen miles long. In Westchester county, New York, there were 2000 cars with us, but they were four abreast, and the procession was not so long in miles.

"The tour may be described as a continuous automobile parade ever since we left Plymouth. It is the longest continuous procession of any kind in all history.

"The route is arranged with the county as a unit, and as we cross each new county line we are met by a new delegation. At the Indiana state line fully 3,000 persons met us and convoyed us to Fort Wayne. An airplane traveled in advance of us across Indiana and distributed notices of the meetings and the time of arrival of the convoy.

"The route of the parade was in many places decorated with flags on nearly every residence and business house, and the flag display was continuous across the country. Persons drove their cars from twenty miles back to see the caravan, and at one crossroads an aged couple were seen displaying a well-used American flag. They had walked three miles to tell us, in the language of the venerable woman: 'Here are two more votes for Coolidge.'

"The tour has averaged about 125 miles per day. We have not traveled Sundays, and the party came in from Evanston today informally to enjoy a rest and a variation from the tour routine in your beautiful city."

The caravan consists of five passenger automobiles and a truck, the latter carrying pictures of the presidential candidates and of the birthplace of the president. Two Dodge cars, a Studebaker and the Reo truck have been with the party all the way from Plymouth. There are two Lincoln cars continuously with the parade as well, but these are arranged by divisions of the Lincoln sales force. The personnel with the caravan consists of the three speakers named, Mr. Cowan, C. T. Doyle, secretary of the tour, A. W. Jefferies, Jr. and William Wallace of Omaha, E. S. Seidel, Ernest Hoch and William Mollden, all of New York.

Mr. Jefferies is no stranger in Utah, having spent three weeks in this state four years ago, and addressed four meetings in 1922. He has spoken on the tour 150 times, of which 125 have been at open-air meetings.

"The thing that strikes me most in this transcontinental trip," said the Nebraska Congressman last night, "is the loss of Davis and the Democratic party. I cannot see but that Davis will run a poor third in the popular vote."

"Every state that we have crossed has been for Coolidge; there does not appear to be any doubt of that. New York will go for him by 1,000,000 plurality. We spent four and one-half days in the northern and western part of that state. All the other states we passed also are safe. Iowa has some LaFollette sentiment, and the situation is somewhat mixed by the Brookhart episode. But Iowa will go Republican. There is some LaFollette vote in Wyoming also."

The presentation of Lincoln voters has been a feature of the tour, and more than 1000 of these have been publically honored. There have been several who voted for Fremont, and one who voted for Winfield Scott, in the days before the Republican party was formed. He was J. S. McElwain of Holyoke, Mass., 96 years old. At Holyoke also were presented three brothers, all of whom had voted for Lincoln. They were R. T. Prentiss and M. W. Prentiss, both of Holyoke, and D. B. Prentiss of South Hadley Falls, respectively, 82, 84 and 86 years old.

Attractive Coolidge Dawes jugate 1-3/4" button with 5-1/2" blue and white ribbon

1924 Coolidge & Dawes campaign poster. 14" x 11"

CAMPAIGN MEMORABILIA IN 1924

Campaign memorabilia for 1924 is relatively scarce. There were several common Coolidge-designed name buttons and perhaps 200 more scare designs of Coolidge buttons, including over 20 Coolidge-Dawes jugate designs (buttons showing the pictures of both president and vice president candidate), the most sought after being a large 6" button. There were a number of campaign brochures, fliers and posters, and a few pennants, ribbons, watch fobs, thimbles, cardboard hand fans, a campaign phonograph record, even bumperstickers!

The Democrats did not spend a lot of money on campaign materials so most Davis items are considered rare, even the Davis name buttons. (Collectors must be careful not to mistake "Davis" name buttons for numerous candidates named "Davis" for local and state offices.) Davis campaign paper items are also scarce.

The anti-Teapot Dome theme and a slogan "Honest Days with Davis" resulted in some interesting buttons but they were less than effective with the voters. Coolidge refused to become personally involved in the Teapot Dome scandal question and the fact that he launched a government investigation helped maintain the people's trust in the President.

The LaFollette-Wheeler campaign also used the Teapot Dome theme on a couple of their campaign items, as well as the liberty bell. The Progressives aimed at a coalition of farmers and workers designed to "break the power of the private monopoly system over the economic and political life of the American people." It is interesting that although LaFollette and Wheeler were the candidates of the Progressive Party, they also were listed as the 1924 candidates of the Socialist Party in a few states, thus appearing twice on some ballots.

Jugates for both Davis and LaFollette are highly sought after with the Davis & Bryan examples usually selling for several thousand dollars.

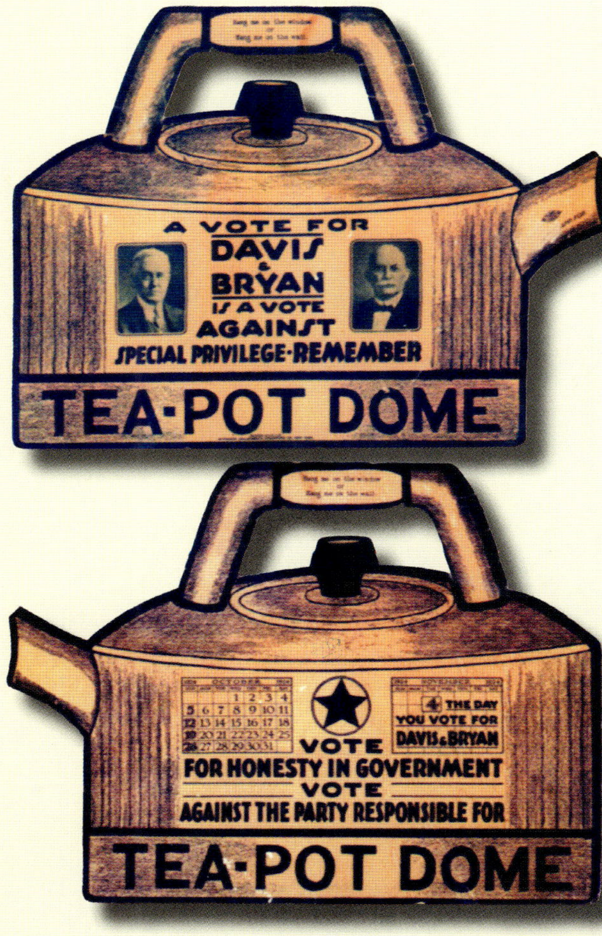

1924 Die-Cut cardboard hanger for Davis & Bryan and exploiting the Teapot Dome Scandal of the Harding Administration, 13" wide

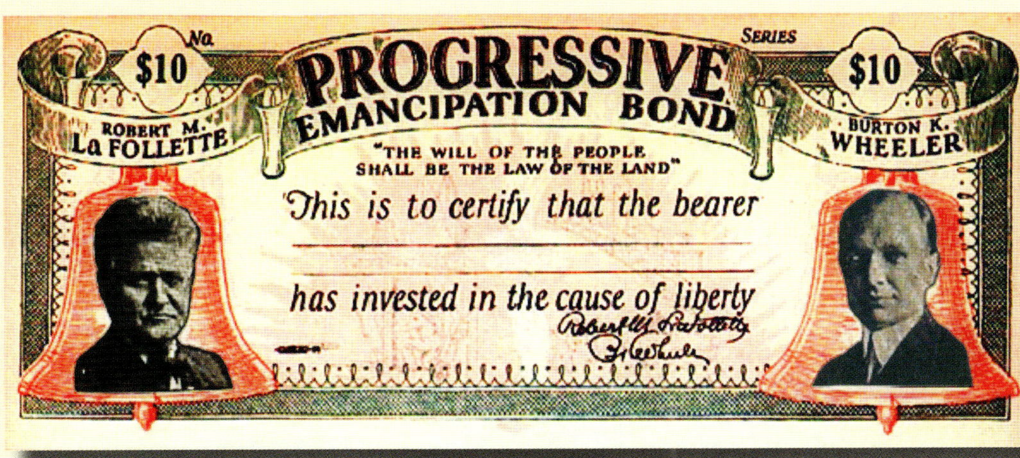

Colorful La Follette-Wheeler Progressive Emancipation Bond - "The Will of the People Shall be the Law of the Land"

1924 Coolidge Dawes campaign metal watch fob suspended from leather strap

1924 Coolidge campaign speech phonograph record

Tuesday, October 14
Day 36

Left Salt Lake City, Utah at 8:00 a.m.

Log: Made 22 stops of 10 minutes each. 246 miles. Spent night at Hotel Atkin, Milford, Utah.

Review: 22 stops today, 10 minutes each, every stop made on scheduled time, average 50 miles an hour between stops.

Pitton driving Lincoln left us at Salt Lake.

We left the Lincoln Highway for the first time today and are now following the Arrowhead Trail to Ely, Nevada.

—E. A. Seidel

Reo Speed Wagon Plays Important Part in Coolidge-Dawes Caravan Touring Continent

REO

Speed Wagon

For light delivery

Dependable and economical

WEETER-COLLETT MOTOR CO.

465 South Main Wasatch 3423

The Reo speed wagon being used in one of the most unique campaign tours in the history of American politics. The caravan is scheduled to arrive in Salt Lake Today.

Presidential campaigns have advanced a long way in the period spanned by three or four administrations, and now, through the use of the automobile, radio and other developments of scientific and industrial progress, candidates for the highest office of the nation can come closer to their constituents than has ever before been possible.

The Coolidge-Dawes caravan, crossing the continent on the Lincoln highway in a little more than a month, carrying the message of the Republican nominees to voters throughout the country, is an example of this. A fleet of four passenger cars and the Reo Speed Wagon pictured here make up the Caravan which left Plymouth, Vt., birthplace of President Coolidge, on September 9, and which is expected to arrive in San Francisco October 20.

One of the remarkable things brought to notice by this caravan is that the automobile industry has evolved a means of transportation such as the Reo Speed Wagon, a car that will assume a capacity load and yet keep pace with passenger cars in the run across the continent. Reports from the caravan en route state that the Speed Wagon is maintaining the required speed all day long, carrying a full load, and running smoothly as a watch. (Adv.)

REO advertisement

Wednesday, October 15
Day 37

Left Milford, Utah at 9:00 a.m.

Log: Stops at Garrison, Baker and Ely, Nevada. 170 miles. Spent night at Great Northern Hotel in Ely.

Review: Crossing desert land now and saw a large eagle today.

Also, lots of skulls and bones of animals.

Garrison is the place where "The Covered Wagon" was filmed.

Ely, Nevada is the home of "Tex" Rickard. Gambling is legal in this state.

Prairie dogs, coyotes and wild horses were seen in abundance. Also a pecular black bird resembling the crow but known as "the Scavenger of the Plains."

—E. A. Seidel

The Ely Daily Times carried the following article:

Campaign Tour in Speedwagon is Unique One

The REO speedwagon which is being used in one of the most unique campaign tours in the history of American politics, namely the Coolidge-Dawes caravan, is scheduled to arrive in Ely today.

Presidential campaigns have advanced a long way in the period spanned by three or four administrations, and now, through the use of the automobile, radio and other developments of scientific and industrial progress, candidates for the highest office of the nation can come closer to their constituents than has ever before been possible.

"Gasing up" somewhere in Nevada

The Coolidge-Dawes caravan, crossing the continent on the Lincoln highway in a little more than a month, carrying the message of the republican nominees to voters throughout the country, is an example of this. A fleet of four passenger cars and the REO speedwagon make up the caravan which left Plymouth, Vermont, birthplace of President Coolidge, on September 9th and which is expected to arrive in San Francisco on October 20th.

One of the remarkable things brought to notice by this caravan is that the automobile industry has evolved a means of transportation such as the REO speedwagon, a car that will assume a capacity load of 3000 pounds and yet keep pace with passenger cars in the run across the continent.

Reports from the caravan in route state that the speedwagon is maintaining the required speed all day long, and running as smoothly as a watch. The Lincoln Highway Garage will be the headquarters of the caravan where the Reo line of cars and trucks are sold."

A brief stop at a Frontier Town

Thursday, October 16
Day 38

Left Ely, Nevada (no specified time)

Log: Drove to Austin, Nevada. 145 miles. Spent night at International House, Austin.

Review: No towns to stop at; all sand and sagebrush.

Republican nominee for Congress, Samuel Arentz, campaigning through Nevada with us.

Austin once had a population of 10,000 people and was one of the "boom" towns in the west. There is now less than 500 people living there, and over $60 million worth of silver taken out of it's hills.

The International House, at which we are stopping, has sheltered some great celebrities including Bret Harte, Teddy Roosevelt and Mark Twain.

—E. A. Seidel

A "desert oasis" - Frenchman's Rest, Nevada with water at 25 cents per gallon and gasoline at 60 cents a gallon

Friday, October 17
Day 39

Left Austin, Nevada at 9:00 a.m.

Log: Drove to Fallon, Nevada. 187 miles. Spent night at Overland Hotel in Fallon.

Review: We left Austin, Nevada this morning in a snowstorm.

Arrived in Fallon at 6:30 p.m.; first paved street since we left Salt Lake.

Duck dinner at the Methodist Church.

—E. A. Seidel

Molden, Moore, Seidel, Doyle, Wallace, Congressman Arentz and Senator Pitt pose for picture at Burnt Cabin Springs, Nevada

The REO Speed Wagon with it's Vermont visuals seems totally out of place at the Burnt Cabin Springs, Nevada stop

Saturday, October 18
Day 40

Left Fallon, Nevada at 9:30 a.m.

Log: Stops at Carson City and Reno. 120 miles. Spent night at Hotel Golden in Reno, Nevada.

Review: Met by Reo dealer at Carson City with eight new cars and escorted into Reno.

Herb Moore, the sage of Plymouth, Vermont, is leaving us here to return to Plymouth, as he is running for election in the state House of Representatives and considers it necessary to be there to get the entire 29 votes, which is the population of Plymouth, Vermont. We mourn our loss!

—E. A. Seidel

Herb Moore leaving the Caravan at Reno, heads back home to campaign in Vermcnt

The October 19 *Reno Evening Gazette* carried an article entitled **"Coolidge Caravan Stops at Carson"**

Carson City – The Coolidge caravan came to town today about noon, from lahontan and Dayton way and held forth at the court-house to about 100 listeners from 2 to 3 o'clock in the afternoon. The visitors had to leave early so as to reach Reno for a 4 o'clock meeting at the university.

Former Governor R. K. Colcord presided and introduced former Congressman Samuel Arentz and the strangers. Mr. Arentz spoke briefly and was cordially received. The other speakers made skill ful appeals, were listened to closely and applauded at intervals. The first speaker, M. W. Hit, resident of President Coolidge's home town of Plymouth, Vermont, told what a simple man the candidate is. The second, Herbert L. Moore, talked about republican economy, and the third, A. W. Jefferis, a former congressman, told his hearers that Coolidge may not be elected because of LaFollette and that the senate may elect a vice-president who will act as president.

A parade marked the entrance of the caravan to town about noontime.

Sunday, October 19
Day 41

Left Reno, Nevada (no specified time)

Log: Drove to Sacramento, California. 146 miles. Spent night at Travelers Hotel in Sacramento.

Review: Crossed the two highest peaks of the California Sierras in the snow.
—E. A. Seidel

The *Reno Evening Gazette*, October 20, carried a report:

Big Crowd Greets G.O.P. Speakers at Rally

Keeping men at work at fair wages is one of the most important things that a Republican congress will accomplish through a protective tariff, Samuel S. Arentz, Republican candidate for representative in Congress, told a large audience which gathered at the rialto Saturday evening to hear Mr. Arentz and members of the Coolidge-Dawes caravan. A well filled house greeted the speakers, despite other attractions of the evening, including the shrine ceremonies, and Mr. Arentz received a big ovation when he was introduced by Mrs. George W. Brady, national Republican committee woman from Nevada, who presided at the meeting.

The other speakers were Herbert L. Moore, neighbor of the Coolidges at Plymouth, Vt.; C. M. Hitt, Iowa farmer and business man, and A. W. Jefferis, Nebraska congressman.

The meeting was opened by W. S. Lunsford, Washoe county Republican chairman, who presented Mrs. Brady.

Mrs. Brady was applauded vigorously when she presented the Nevada state flag to Congressman A. W. Jefferis, who, upon his return to Washington, will deliver it to President Coolidge as a token from the Republicans of Nevada. As Nevada, the battle-horn state saved the nation in '64, her three electoral votes may save the nation in 1924, Mrs. Brady said amid cheers.

Mr. Moore explained that the caravan idea originated among the neighbors of the president at Plymouth who became aroused when it was insinuated by LaFollette speakers that Mr. Coolidge never had been a real farmer. "We concluded to go out and refute this statement to the world," he said.

Mr. Hitt warned against any such change in the Constitution as that proposed by LaFollette to curtail the power of the supreme court. "The supreme court is the people's umpire," he said. "The Constitution of the United States guarantees certain liberties and rights to the individuals of the nation and the supreme court functions to protect the people against any possible oppression by congress." Mr. Hitt characterized the LaFollette appeal as a dangerous doctrine.

Congressman Jefferis closed with a fifteen-minute review of the accomplishments of the administration, stressing the great reduction in the public debt and national expense which had been effected through the budget system and other methods since President Harding was elected.

This California item - Keep Cool with Coolidge. Cal. for "Cal" - is a heavy canvas automobile tire cover for the "spare tire" which often was prominently displayed on the rear of the vehicle. 32" diameter

Monday, October 20
Day 42

Left Sacramento, California at 10:00 a.m.

Log: Stops at Lodi, Stockton, Modesto, Turlock, Merced, Madera and Fresno, California. 180 miles. Spent night at Hotel Fresno in Fresno.

Review: All paved roads through here and beautiful scenery. got our first look at the palm trees and the fruit orchards.

C. S. Nussbaum joined us at Sacramento as an added speaker. He will campaign with us through California.

C. M. Hellbach of Frisco now driving the Lincoln.

Good, snappy motor-cycle escort all the way.

—E. A. Seidel

The *Lodi Sentinel* reported the following announcement:

CAMPAIGN FOR COOLIDGE WILL START IN LODI

Organization of the general county campaign committee for Coolidge and Dawes is under way, indications being that the body will consist of appropriately 200 members representative of all sections of San Joaquin county.

Each member of the San Joaquin county Republican committee has been requested to name a certain number of persons from his district for the campaign committee, and these nominations are now being received at headquarters.

For the next Coolidge-Dawes meeting in Stockton, the date for which is to be set later, Senator Lonroot and Senator Samuel Shortridge are being sought as speakers. Senator Lonroot will be on the coast this month and efforts to have him visit here are being extended. Appearance of Senator Shortridge is practically assured.

Coolidge meetings throughout the county will be commenced next week with sessions in Ripon and Lodi, and possibly other places. A. F. Roberts in the Linden section and O. I. Marrs in Lodi are planning for campaign meetings. Coolidge stickers are now being received and will make their appearance on automobiles soon.

On October 19 the Coolidge-Dawes Plymouth caravan will visit this section. This is an auto caravan from the home town of President Coolidge, and it is touring the country telling the people of the boyhood of Calvin Coolidge and giving intimate sketches of the man, a real closeup of Calvin Coolidge, the man and fellow townsman.

The caravan will visit forty-five cities in California, and will then tour Washington and Oregon. The caravan left Plymouth September 9, and during the first half of the journey has covered ten states, with more than 200 stops in cities on the Lincoln highway. The caravan is in charge of Hon. A. W. Jeffries of Omaha, former congressman from Nebraska, who notified Gen. Charles W. Dawes of his nomination as vice presidential candidate. The party also includes former Congressman M. W. Pitt of Iowa and Herbert Moore, a boyhood chum of the President.

A. V. Faight, who is in charge of Republican headquarters at 20 North San Joaquin street, Stockton, reports many visitors to headquarters and keen interest in the Republican campaign. If the intrest shown by the many visitors to headquarters is an indication of the sentiment of the county, Coolidge and Dawes will carry San Joaquin by handsome majorities, says Mr. Faight. The sentiment among the farmers particularly is strong for the Republican candidates and many are asking for Coolidge stickers, buttons and literature.

Visitors are always welcome at headquarters and all Republicans are invited to drop in with suggestions, advice, information or to receive any suggestions or information desired.

"Fly-Cops" California motorcycle police escort

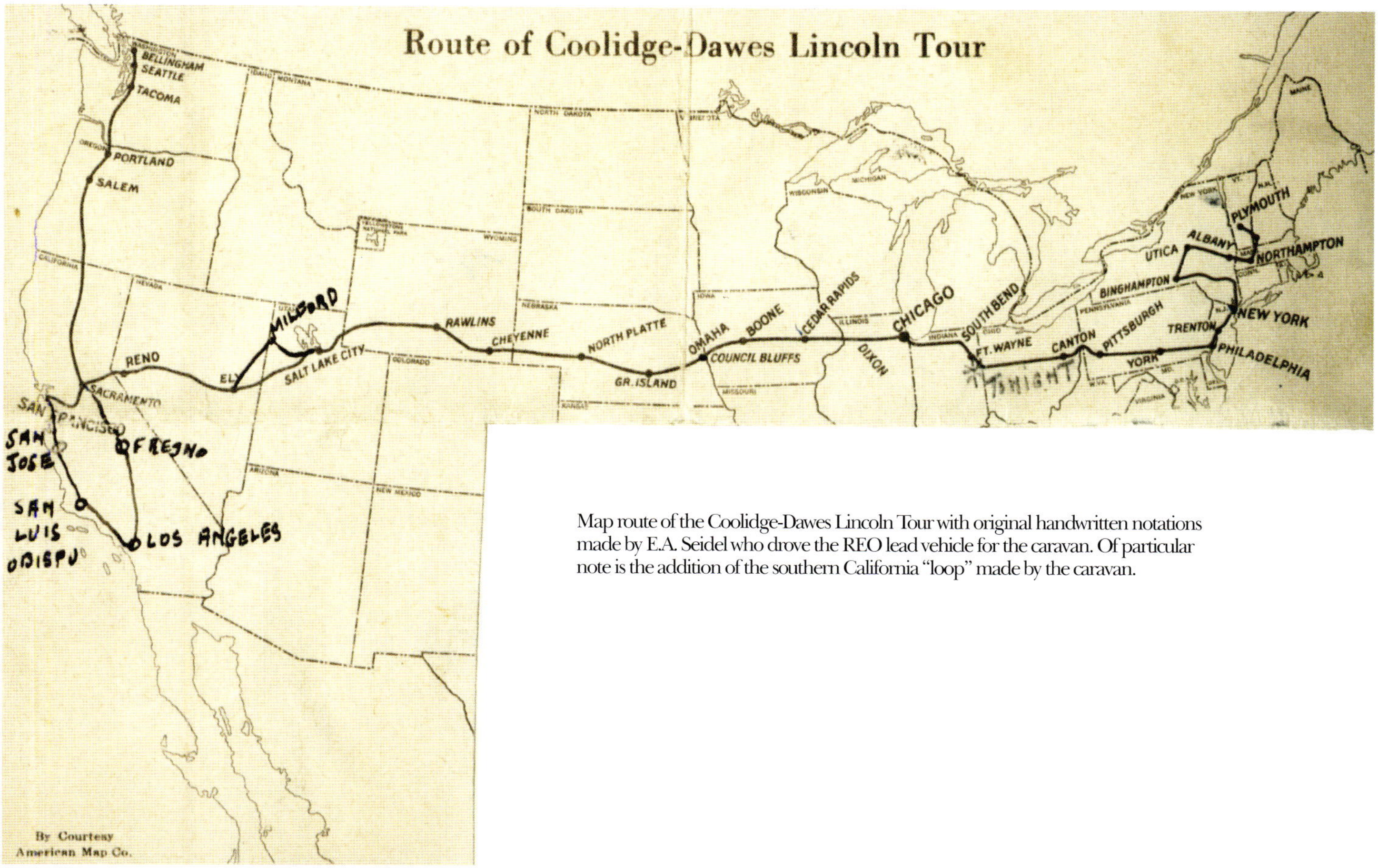

Map route of the Coolidge-Dawes Lincoln Tour with original handwritten notations made by E.A. Seidel who drove the REO lead vehicle for the caravan. Of particular note is the addition of the southern California "loop" made by the caravan.

Tuesday, October 21
Day 43

Left Fresno, California at 7:00 a.m.

Log: Stops at Bakersfield, Hollywood and Los Angeles. 249 miles. Spent night at Hotel Rosslyn in Los Angeles.

Escort: Met by B. C. Foy, the Reo dealer.

Review: Made first 114 miles between Fresno and Bakersfield in 2:28 minutes without pace.

C. L. Weilti of Salt Lake City left us here; also Senator Pitt, who received a telegram that his aged mother was dying.

Had dinner at the University Club, and saw Harry Carroll's new revue, "Pickings" at the Orange Grove. Heading the cast were New York's own favorites including Will Morrisey, Midgie Miller and Eddie Borden.

—E. A. Seidel

The October 21 front page of *The Bakersfield Californian* carried the following feature:

SPEAKER ADDRESSES STREET CROWN FOR COOLIDGE-DAWES.

Preceded by a truck covered with the dust of many states which arrived early this morning to make preparations for the automobile train, the Coolidge-Dawes Plymouth caravan, on tour over the Lincoln Highway from Plymouth, VT to the Pacific coast, arrived in Bakersfield shortly before noon today for a short stop to campaign for the Republican candidates.

The caravan itself was met by members of the Kern County Coolidge-Dawes league on the Fresno highway near Beardsley and was escorted into the city to the Southern hotel, where a stop was made for a short open air meeting. Several prominent Republicans, on the tour from Plymouth, President Calvin Coolidge's birthplace, made talks at the meeting.

With the bear flag as an acknowledgment to the greeting given the caravan upon its arrival in California, the fifteenth state flag to be presented the travelers, in his hand. M. W. Pitt, former speaker of the Iowa house of representatives, talked to the crowd of approximately 200 persons that assembled in front of the Southern hotel.

Mr. Pitt said that again the country is at a crossroad, as it has been in the past 34 years and that its people are again asked to detour. The speaker continued by stating that each time the country has detoured that it has asked to come back to the main road of the Republican party by an almost unanimous vote. He said that an attack upon LaFollette's following, which he says includes the radicals of all descriptions.

The other speaker was A. W. Jefferies, director of the caravan, who spoke upon Coolidge and his history from boyhood and also upon the cost of government. He urged that all should take an interest in the campaign as all must pay taxes indirectly if not directly.

He used a pair of shoes as an example of indirect taxation, saying that there is a tax paid by the packer that gets the cow and sells the hide, still further taxes by the tanner, manufacturer, wholesaler and retailer. The second speaker claimed that taxes are being reduced under Republican legislation.

Herbert L. Moore, a farmer of Plymouth and boyhood friend of President Coolidge, left the caravan at Reno, Nev. The other members of the party, besides the two speakers, were John P. Cowan of Pittsburgh, PA, manager; Frank Winch of Wilmington, Dela, advance representative, E. S. Seidel, Ernest Hoch and S. E. Moliden, New York and William Wallace and A. W. Jefferies, Jr., Omaha, drivers.

5,000,000 see Party

It is estimated by Mr. Cowan that more than 100,000 automobiles have participated in the motor escort provided from county to county across 15 states and that 5,000,000 people have witnessed the progress of the campaign parade

The party continued upon its journey northward at 12:30 o'clock. Among those welcoming the visitors were James Pytrini, Dr. Fred J. Crease, C. K. Badger, and other members of the local Coolidge-Dawes league.

This 6" colorful Coolidge-Dawes jugate button is undoubtedly considered one of the very best Coolidge items of the 1924 campaign

The *Los Angeles Times* carried the following October 21 feature:

CARAVAN IN SAN JOAQUIN.

Coolidge Band Gives Message of Tariff on Way to Huge Rally Tonight in Los Angeles.

Through the heard of the San Joaquin Valley the Coolidge-Dawes Caravan sped today carrying to thousands of Californian farmers, from Sacramento to this city, the real meaning of protective tariff in its relation to the prosperity of this State, and told how the future of the country is being menaced by the sovietic teachings of Robert M. LaFollette.

Enthusiastic was the greeting accorded the Pilgrims from the Green Mountain State and every town along the route heard the message of "Americanism and Common Sense," carried by speakers who started with the caravan from Plymouth, Vt., the birthplace of President Coolidge.

GIVEN STATE FLAG

Before leaving Sacramento for its pilgrimage through California's rich agricultural valley, the leaders of the Coolidge-Dawes caravan were presented with the State flag of California, by members of the Coolidge Club of that city. Lester J. Hinsdale presented the flag, and told the crusaders from New England to return to Vermont with the glad tidings that California was going to give Coolidge a 200,000 majority.

At Stockton, a monster meeting was staged in front of the County Courthouse. The meeting was opened by Charles L. Neumiller, chairman of the State Central Republican Committee. Nelson H. Pitt, former Speaker of the House of Representatives of the Iowa Legislature, was the first speaker and called attention to the general prosperity now being enjoyed by California farmers and workers. He dwelt at length upon the possibility of this country being overrun with the products of cheap foreign labor, if the principles advocated by LaFollette should overthrow the present tariff.

"Today American labor is receiving the highest wage in the world," said Pitt, "but remember there is a new hound on the track that would put an end to these things – so labor must be on its guard. Another thing, and that is the individual liberty of every American is now being menaced by the teachings of LaFollette."

NO MORE SOUP LINES

A. W. Jeffries, former Congressman from Nebraska, recalled both at Stockton and Fresno the achievements of the Republican administration since March 1921.

"Remember when the present Republican administration took office," said the Nebraskan, "that there were nearly, 5,000,000 men in this country idle, and we were going through the 'soup kitchen' era as this country had done in 1893, but today what have you? Practically no unemployment in this country, and the savings annually to the taxpayers of about $2,000,000,000 through the budget system. The Republican administration is making possible the collection of the European war debt, and that means for the farmer lowered taxes. One of the reasons this San Joaquin Valley should lend its every effort to maintain the Republican administration in office is because it has, and will continue to do so, protected the farmers of California with the protective tariff."

OUTPOURING OF VOTERS

Fresno was the scene tonight of one of the largest outpouring of voters since the formation of the Coolidge-Dawes campaign forces in this city.

Capt. C. F. Nusbaum, formerly of Kansas, and a recent resident of Sacramento, told the farmers and industrial classes of this city just what LaFollette meant for them in the way of lowered wages, and an invasion of their individual liberties

Tomorrow morning, with the break of dawn, the Coolidge-Dawes caravan leaves for Bakersfield, where a large outdoor noon rally is to be held, at the conclusion of which the party will leave for Los Angeles.

RALLY TONIGHT FOR COOLIDGE CARAVAN

Closing with a great rally tonight at Trinity Auditorium, the Coolidge-Dawes Plymouth Caravan, which has toured the Lincoln Highway across the country from the birthplace of the President, will proceed through the San Fernando Valley to Los Angeles this afternoon.

The principal speaker at the Trinity Auditorium will be United States Senator Samuel Shortridge, who has been delivering addresses throughout California in the interests of the Republican ticket. Other nationally known orators who will speak are: A. W. Jeffries, former Representative from Nebraska; M. W. Pitt, former Representative from Iowa, and Herbert Moore, a boyhood friend of the President, all of whom are with the caravan.

The object of the caravan is to acquaint the people with the true character of the President by revealing incidents of his youth and manhood on the Vermont farm and to portray the principles to which he stands.

VISITS 200 CITIES

The caravan left Plymouth, Vt, the birthplace of the President on September 9, where it was originated by the Plymouth Coolidge Club and has toured across the United States, visiting more than 200 cities. Forty-five California cities will be reached by the caravan.

Stops will be made by the caravan this afternoon at Saugus, Burbank, Glendale and Pasadena and Los Angeles will be entered early in the evening. The meeting at the Trinity Auditorium will be preceded by a torchlight procession, starting at the north end of the Broadway tunnel at 7 p.m.

Several hundred automobiles bearing Coolidge enthusiasts will assemble at the Coolidge-Dawes headquarters, 220 South Brand Boulevard, Glendale, and at 3 p.m. will proceed to Saugus, there to welcome the caravan to Los Angeles county. Each Coolidge-Dawes Club in the county will be represented, Pasadena having promised to send twenty-five cars.

BANNERS FOR CARS

Banners and other decorations will be distributed at the Glendale headquarters to those cars not already bearing them. Senator Shortridge and other notables will be present in the escort. At a signal from the band the escort will start for Saugus.

The escort and the caravan will meet at Saugus at 4:30 p.m., where the caravan will be welcomed by the chairmen of the various committees of Southern California. After a short ceremony at Saugus, the procession will depart for Burbank, where a number of speeches will be delivered.

At Pasadena the time for the departure will be announced, and an assembly will be made at the Broadway tunnel at 7 p.m., where additional automobiles will take their place in the line. A parade will then be formed, the route being along Broadway to Ninth street, west on Ninth street to the Trinity Auditorium.

NORTH BY COAST ROUTE

Band music, an organ recital and singing will precede the speeches of the caravan orators and Senator Shortridge.

The caravan arrived at Colfax last Saturday, where it was greeted by a delegation of Californians, who escorted it to Sacramento. After a brief stop there the caravan left for Southern California by way of the San Joaquin Valley, where halts were made at numerous cities. A great wave of enthusiasm has followed the caravan, according to advices, during the three days it has been in California.

Following tonight's rally, the caravan will turn north, going to San Francisco by way of the Coast Route. From San Francisco it will continue to the States of Oregon and Washington.

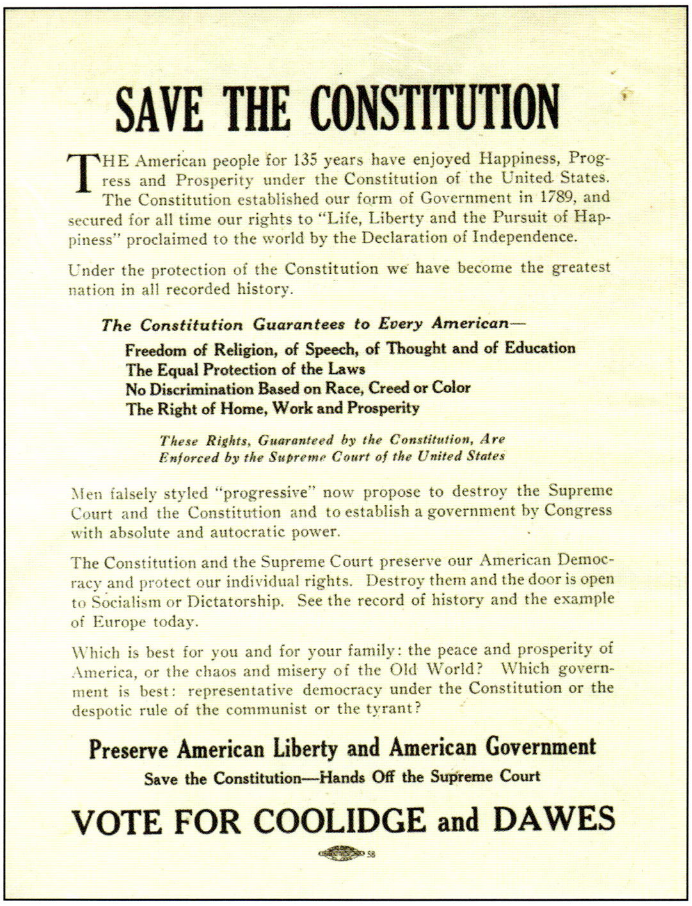

SAVE THE CONSTITUTION

THE American people for 135 years have enjoyed Happiness, Progress and Prosperity under the Constitution of the United States. The Constitution established our form of Government in 1789, and secured for all time our rights to "Life, Liberty and the Pursuit of Happiness" proclaimed to the world by the Declaration of Independence.

Under the protection of the Constitution we have become the greatest nation in all recorded history.

The Constitution Guarantees to Every American—

Freedom of Religion, of Speech, of Thought and of Education
The Equal Protection of the Laws
No Discrimination Based on Race, Creed or Color
The Right of Home, Work and Prosperity

These Rights, Guaranteed by the Constitution, Are Enforced by the Supreme Court of the United States

Men falsely styled "progressive" now propose to destroy the Supreme Court and the Constitution and to establish a government by Congress with absolute and autocratic power.

The Constitution and the Supreme Court preserve our American Democracy and protect our individual rights. Destroy them and the door is open to Socialism or Dictatorship. See the record of history and the example of Europe today.

Which is best for you and for your family: the peace and prosperity of America, or the chaos and misery of the Old World? Which government is best: representative democracy under the Constitution or the despotic rule of the communist or the tyrant?

Preserve American Liberty and American Government

Save the Constitution—Hands Off the Supreme Court

VOTE FOR COOLIDGE and DAWES

"Save the Constitution" was one of the primary themes of the Coolidge campaign, heard over and over during the speeches of the Coolidge-Dawes Lincoln Tour and targeted primarily at the La Follette-Wheeler ticket

Unusual "Coolidge for President" engraved 2" celluloid disc with nine inset facsimile rubies and holed at top - probably designed with the women voters in mind

Wednesday, October 22
Day 44

Left Los Angeles, California at 11:00 a.m.

Log: Stops at Ventura, Carpentica, Santa Barbara, Orcutt, Santa Maria and San Luis Obispo. 247 miles. Spent night at Hotel Andrews in San Luis Obispo.

Review: Roads conditions wonderful; weather very warm. Road running for 50 miles along side the Pacific Ocean.
—E. A. Seidel

The October 22 *Ventura Daily Post* ran a brief front page story:

REPUBLICAN CARAVAN HERE THIS MORNING

Big Public Meeting with President's Friends and Neighbors as Speakers, to be Held Here

Republicans, attention!

Every Republican who has a car is urged to go to the city limits at the forks of the road this morning at 10 o'clock to meet the Coolidge-Dawes caravan which will come here from Los Angeles arriving at that hour at the forks of the road here.

The caravan which is a trans-continental tour, will hold a mass meeting at Main and California streets this morning at 10:30 o'clock. There will be speakers of national note present and everyone is urged to hear them.

This caravan started from Plymouth, Vt., President Coolidge's home, on September 9. Their destination is San Francisco, where they are due October 18.

The idea of the tour was suggested by some Vermont friends and neighbors of President Coolidge, some of whom are in the party."

In addition to buttons, campaign position leaflets were popular tools used by all of the 1924 candidates

Calvin Coolidge scarce campaign bank supports the theme of Coolidge being a fiscal conservative. 4-1/4" high

Crowds turn out in Santa Barbara, California

The *Ventura Free Press*, which comes out later in the day, covered the event in great detail as their front page lead story:

GOVERNMENT IS SAFE IN "CAL'S" HANDS

Caravan Speaker Says He Will Give People Sound, Practical Administration For Next Four Years.

"This government is safe in the hands of Calvin Coolidge," declared H. F. Nusbaum of San Francisco, speaking here for a few minutes this morning on the corner of Main and California streets as a member of the Coolidge Lincoln Tour Caravan, "and it behooves the people of these United States to return him to office if they want a sound, practical administration with a continuation of the country's prosperity."

He flayed LaFollette, declaring him not a Progressive as he claims but an out and out Socialist. He said he was a dictator, and that if he could not rule, his policy was to ruin. He declared that LaFollette will not get as many votes on November 4 as he would have polled four weeks ago.

The caravan did not stop but a few minutes here owing to having to make 20 miles today. There were six cars in the cavalcade, and 15 people.

One of the features of today's meeting here was the presence of two of the oldest voters in California.

They were Dr. S. L. Stewart and E. P. Foster, both of whom cast their first ballot for Abraham Lincoln in '61.

The caravan today is beginning the last leg of its tour to the Golden Gate. This has been the longest procession in history and it is the first instance that a trans-continental trip has been made by motor vehicles on a fixed schedule and arrived on time.

More than 100,000 automobiles have participated in the motor escort provided from county to county in the long journey across the country. The caravan has traveled 15 states and 250 meetings have been held with an aggregate attendance of a quarter of a million voters. It is estimated that 5,000,000 people have witnessed the progress of this unique campaign speaking parade, taking part in the receptions or in cheering the travelers on their way.

The little caravan that left the birthplace of President Coolidge in the Green Mountains of Vermont has been kept intact, the, the six motor cars and the same drivers making the entire journey to the coast, but at times hundred of cars followed and in some places the procession was from 10 to 15 miles in length. The journey was begun at noon on Tuesday, September 9, and the trip has required 36 days' driving time, the daily average mileage being about 111 miles. The tourists encountered all kinds of weather but every day was made on schedule and the only delays encountered were due to heavy traffic in some of the eastern states.

The route covered southern Vermont and New Hampshire, western Massachusetts and central New York, and after leaving New York city followed the route of the Lincoln highway across the continent. More than a dozen mountain ranges were crossed, including the Berkshires and Catskills, the Alleghanies, the Rockies and the Sierras. In making their way through Wyoming a lot of snow on the Laramie range failed to baffle the intrepid drivers of the cars. In one mountain community an escort of cowboys on ponies led the way for the mud-covered automobiles.

The caravan carries the state flags of 15 states presented by the women's committees of each state crossed on the way. These will be preserved as mementoes of the journey by the Republican national committee.

Directly beside the front page article in the *Ventura Free Press* was an editorial endorsement for Coolidge:

Ventura County Must Be For Coolidge

Because of its great lemon industry, because of its great dried fruit production, because of its wonderful walnut industry, because of its beans and its beets and its almonds, Ventura county is almost under absolute obligation to support Calvin Coolidge in the coming election for president of the United States.

Without protection Ventura County's industries will suffer and our land owners will not receive a just return on their investment.

We must not be unmindful of the fact that under the law as it now exists the president of the United States and not Congress fixes and hereafter will fix the rates of duty to protect our industries. Do we want as president a man who declares he will grade them down or abolish them entirely, or a man who is pledged with his party to maintain them at a mark that will stimulate our development, encourage our industrial progress and continue or condition of prosperity?

Sixty thousand California farmers petitioned Congress to increase the duty on almonds to protect that industry from the cheap labor of the Mediterranean countries. LaFollette voted to reduce the duty and to put some almond products on the free list. The Republican majority heeded the appeal of the California growers.

The records show that LaFollette held the same attitude toward our citrus industry, our olives, our walnuts, our rice, our beans, our barley, our sugar beets, our dried and preserved fruits, our poultry and a hundred other products of California soil. These industries, however, are protected because the Republican party and President Coolidge were not unmindful of California's appeals.

Do Ventura County farmers want protection for their products or do they want to take a dose of hard times that LaFollette does not deny he has prepared for them?

Is it not necessary for the welfare of Ventura County to keep Coolidge on the job?

The San Luis Obispo newspaper ran the following account on the Coolidge-Dawes caravan visit:

Coolidge-Dawes Caravan Stops Here Overnight

After having come all the way from Plymouth, Vt., the birthplace of Calvin C. Coolidge, to carry the message of his neighbors and townspeople to the voters of the nation, the Coolidge-Dawes Lincoln caravan reached this city last night. The members of the caravan spent the night at the Andrews and will resume their journey this morning, stopping tonight at San Jose. The caravan is now northward bound, with Seattle as its ultimate destination.

A. W. Jefferis of Omaha, former Congressman, who nominated Dawes, was the speaker last night at the meeting held by the caravan at the corner of Morro and Monterey streets. He traced the history of Republican administrations and of the reflection of their sound governmental policies in the general prosperity of the country and the economical conduct of public affairs. He contrasted these eras of Republicanism with the periods of panic, hard times and extravagance when the Democratic party was in control.

Need Republican Congress

The speaker emphasized the urgency of electing a Republican Congress to support the administration. He declared that there was no middle ground between right and wrong, or between the wisdom of Republican politics and the theories and follies of Democracy. A third party he compared to a fifth wheel to a vehicle, and if supported as a protest it could only result in confusion and chaos. Its only aim would be to tear down, whereas the election of Coolidge and Dawes meant confidence to the people and the continuation of a constructive policy. He emphasized the importance of the tariff question, especially as it affects California.

West Supports Coolidge

Mr. Jefferis declared that the western states through which the caravan has passed would give decisive support to the Republican candidates. He urged Republicans to stand loyally to the traditions of their party and predicted that a normal Republican turn out of the voters on election day would result in a triumph equal to 1920.

The caravan has traveled more than 4000 miles on its trip and is now on its last leg of the tour to Seattle. This has been the longest procession in history and is the first instance that a trans-continental trip has been made by motor vehicles on a fixed schedule and arrived on time.

The officers of the tour are: Hon. A. W. Jefferis, Omaha, Neb., director; John P. Cowan, Pittsburgh, Pa., manager; Charoes T. Doyle, Pittsburgh, Pa., secretary; and Frank Winch, Wilmington, Del., advance representative. The drivers of the cars, all graduate engineers, are: E. S. Seidel, Ernest Hoch and S. E. Mollden, New York and William Wallace and A. W. Jefferis, Jr., Omaha.

Pitt Leaves Caravan

Mr. Jefferis is the principal speaker on the trip as well as the director of the caravan. Hon. M. W. Pitt, former speaker of the Iowa House of Representatives, who was one of the speakers, received a message in Los Angeles yesterday that his aged mother was dying, and hastily left to be at her bedside.

More than 100,000 automobiles have participated in the motor escort provided from county to county in the long journey across the country. The caravan has traversed 15 states and 250 meetings have been held with an aggregate attendance of a quarter of a million voters. It is estimated that 5,000,000 persons have witnessed the progress of this unique campaign speaking parade, taking part in the receptions or in cheering the travelers on their way.

Trip Begun Sept. 9

The little caravan that left the birthplace of President Coolidge in the Green mountains of Vermont has been kept intact, the six motor cars and the same drivers making the entire journey to the coast, but at times hundreds of cars followed in in some places the procession was from 10 to 15 miles in length. The journey was begun at noon on Tuesday, Sept. 9, and the trip has required 36 days driving time, the daily average mileage being about 111 miles. The tourists encountered all kinds of weather but every day was made on schedule and the only delays encountered were due to heavy traffic in some of the eastern states.

Lincoln Voters

When the caravan departed from Plymouth a group of aged men who had voted for Abraham Lincoln and who knew Calvin Coolidge as a boy wrote their names in a book and presented it to the travelers, asking that all the men found in route who likewise had voted for the first Republican president be enrolled, the object being to give the register of Lincoln voters to President Coolidge. Before the caravan reached Indiana so many Lincoln voters had enrolled that it was necessary to add another volume to the register. In the books are names of scores of men who had voted for Lincoln both in 1860 and in 1864 and there were about a dozen names of men who had voted for John C. Fremont, the first Republican candidate for president in 1856. S. E. McElwain, of Holyoke, Mass., aged 96 years, not only has cast his ballot for every Republican presidential candidate but enjoys the distinction of having voted for Gen. Winfield S. Scott, the last Whig candidate in 1852.

Thursday, October 23
Day 45

Left San Luis Obispo, California at 8:00 a.m.

Log: Stops at Watsonville, Santa Clara, Campbell and San Jose. 202 miles. Spent night at Hotel Vendome in San Jose. California.

Review: Presented with baskets of California grapes at Watsonville.

Dinner given at Commercial Club by War Veterans.

—E. A. Seidel

The *San Jose Mercury Herald* gave the caravan front page billing in their October 24 issue:

COUNTY T'RNS OUT IN FORCE TO GIVE CARAV'N WELCOME.

Man Who Nominated Dawes at Cleveland Convention speaks at Los Gatos.

Commercial Club Filled.

Santa Clara County turned out in force yesterday afternoon to witness the dust-covered Coolidge-Dawes Lincoln Caravan which has covered 6,000 miles in its journey over the Lincoln Highway from Plymouth, Vermont and more than a mile of closely packed automobiles swung into line, and led by eight decorated Lincoln cars, carrying prominent members of the Republican county central and auxillary committees, G.A.R. and World War veterans, followed the caravan from Los Gatos through Campbell and Santa Clara to San Jose.

The caravan and its honor escort from the county, which met it at Watsonville, and piloted it to Santa Cruz and back to Los Gatos, was met at Summit, the county line, by Capt. John Pacheco, Carl Balch, and "Sunny Jim" Margason of the county traffic squad, who held up all traffic to allow the parade right of way. The work of the traffic squad, which sped ahead of the procession and held up traffic at intersecting roads – even holding a Southern Pacific train at Vanona Junction until the caravan and its escort had crossed the railroad tracks – was highly commended by members of the caravan party.

Met By Scout Car

Owing to the efficient work of the scout car, the caravan arrived on scheduled time in Los Gatos, where a big local committee welcomed the party and presented to each member a delicious basket of fruit.

Mr. Jeffries, former congressman from Nebraska, was the first speaker. Although Mr. Jeffries, during his 7-weeks tour with the caravan, had spoken daily from three to 15 times, his voice has lost none of its carrying power and resonance, and for 10 minutes he held the crowd of more than 1,000 gathered in front of the Lynden hotel.

Will Lake's band, in a profusely decorated truck, provided music at Los Gatos and preceded the parade on its San Jose-ward journey, apprising Campbell and Santa Clara that "Hail, Hail the Gangs All Here," as the mile-long procession passed through the towns.

Crowd Hears Coolidge Speech

Every seat at the tables in the Commercial Club, where the caravan members were banqueted, were filled, and both dining rooms were filled to overflowing with San Joseans and Republicans from outside communities who gathered to hear President Coolidge's speech on his plan for tax reduction over the especially installed radio apparatus installed by Ray Newby at the Commercial club. So great was the crowd that four machines were required to allow the big assemblage to hear the speech. Two of these were provided by Mr. Newby, and two were hastily set up by Mr. Lawrence of the Wiley B. Allen company, one in the Commercial club, the other in the concert room of the Allen company.

Because of the fact that the Republican association of war veterans throughout the state were holding joint banquets in their home cities last evening, Walter L. Chrisman, chairman of the Republican county central committee, names Ernest E. Williams, president of the local association, presiding officer of the evening and Mr. Williams in a graceful introduction, presented the speakers of the evening.

Praises County Hospitality

Mr. A. W. Jeffries, an intimate friend of both Coolidge and Dawes, and director of the tour, expressed his pleasure at the cordial reception of Santa Clara county which he said, in enthusiasm and efficiency, exceeded anything the party had received during its California tour.

"We have traveled 6,000 miles, but there has been only one San Jose on the trip," said Mr. Jeffries. "The beauty of the scen-

Hotel Vendome, San Jose, California

ery which we have passed this afternoon will be remembered and cherished as one of the most beautiful memories it has been our privilege to observe, and your enthusiastic welcome and splendid hospitality will never be forgotten."

Mr. Jeffries entertainingly sketched the mission of the caravan from its leaving Plymouth, Vermont on September 9, through the many states through which they have motored, predicting a Coolidge victory on November 4.

New York for Coolidge

"The New England States know Coolidge and are strong for him, and New York is safe for Republicanism this year by more than a million plurality," prophesied Mr. Jeffries. "Nebraska, despite the Bryans, is going for Coolidge and Dawes," he added positively. Wyoming, Utah and the middle west showed their sentiment to be unmistakably for Coolidge, he stated. "Nevada, it has been said, is doubtful," continued Mr. Jeffries. "I don't believe it. Its people showed too much enthusiasm and activity for Coolidge and I believe the stats will cast a big ballot for the Republican candidates."

In Pennsylvania the miners, at first apathetic, were stirred and interested at his statements concerning the tariff and what it means to them. "Coolidge is for economy, and then more economy, and when through economy and the administration of a budget system Coolidge has reduced the cost of government from $5 billion to $3 billion. It means $2 billion rests in the pockets of the people instead of going out in taxes."

"That is why this country is coming out of the mess of war. That is why it means something to the miners, to the farmers, and the workers, and everyone else in the nation. The savings means an annual savings of more than $135 million dollars in interest alone. Taxation has been reduced one and one-quarter billion dollars annually or more than 60 percent in three years. During the present fiscal year taxes still show a savings to the people of approximately 3-1/2 million dollars a day compared with 1921."

Republican Economy

"The $25 billion dollar indebtedness of the country has been reduced to $21 billion dollars and during this time the government has appropriated $250 million dollars in an effort to take care of the men who held our flag in the air during the world war. Liberty bonds, which were selling at 30 and $1 are over par now."

Speaking of LaFollette's proposal to nationalize public utilities, Mr. Jaffries said, "Thirty billion dollars would have to be raised if the government took over the railroads, which would be a tax on every farmer, every laborer, every tax payer in the United States. Of the LaFollette proposal that Congress be given the power to override the Supreme Court, he said: "The day that power is placed in Congress, America will be where it was before the revolutionary war, and the rights of the individual will cease to exist. The poor man will be the first to feel the heel of autocracy."

Rouse Voters to Menace

"It is time for the American people to be aroused to the menace to their liberties. Because of this Coolidge and Dawes allowed this caravan to start forth with its message – because they want to preserve and to conserve the things that make this nation what it is. It is because of the Intelligent, clear vision of Charles Dawes that this nation has performed a service affecting the peace of the world."

"When the reparations commission took his plan back to their people they accepted it voluntarily instead of by force, and thus for the first time in six years, the business of the world is annulled and the Republican party, through whose effort this has come, has laid the foundation for world peace."

"There is a crisis in our history today. It is not the crisis of war, but one calling for loyalty in the time of peace. The time to act is here. Do not let California waver, preserve our constitution, our flag and the opportunity for future generations to enjoy the rights we have today. Preserve American liberty, American industry and American homes by casting your ballot for the man who stands for all these things – Calvin Coolidge."

State Democrats for Coolidge

Capt. C. S. Nusbaum, from Coolidge state headquarters, was the second speaker of the evening. "I am proud of the citizens of California," said Captain Nusbaum, who has been speaking throughout the state. "Everywhere I have gone the leading Democrats have said: 'We cannot afford to let the Socialists and LaFollette take over our government and destroy our liberty and our individualism. We will join the Republicans this year and vote for Calvin Coolidge 100,000 to 150,000 majority,'" he added. "Republicans and Democrats are uniting to strike such a blow to socialism that they will never bob up again. They resent the fact that LaFollette has called American soldiers and American people boobs. Naturally they resent such a slander, and they will show their sanity and rebuke the man who so names them by voting to retain their government, their homes and their property."

Newspaperman Conceived Tour

J. W. Cowan, of Pittsburgh, who conceived the idea of the Coolidge-Dawes caravan, Mrs. Cowan, and their young son, were with the caravan which remained last night at a local hotel. The caravan leaves this morning at 9 o'clock visiting Palo Alto, Redwood City and Burlingame on its way to San Francisco, where they will be welcomed by Mayor Rolph and a reception committee of prominent Republicans.

Friday, October 24
Day 46

Left San Jose, California at 8:30 a.m.

Log: Stops at Palo Alto, Redwood City, Burlingame, San Francisco, Oakland, Vallejo and Sacramento. 233 miles. Spent night at Travellers Hotel in Sacramento.

Review: Big reception at Frisco, headed by the Reo dealer.

Was introduced to Mayor Rolph and had quite a talk. Had our pictures taken together.

Luncheon given by Republican Club at Hotel St. Francis.

Presented with violets and olives at Oakland.

—E. A. Seidel.

Both the ***San Francisco Examiner*** and the ***Illustrated Daily Herald*** gave the caravan's visit number one front page stories with top headlines and multiple photos.

Mayor Gives Welcome at Civic Center. Importance of Maintaining Tariff Stressed by Speakers. Boys of '61 Sign Up. workingmen of S.F. Will Support President, Declares Rolph

Under sunny skies and true San Francisco warmth, there was welcomed at the Civic Center yesterday the Coolidge-Dawes Plymouth Caravan, which started forty-five days ago from the birthplace of President Coolidge, carrying the message of "Common Sense and Americanism."

The caravan headed by a motor truck bearing pictures of the humble home from which the Nation's leader came, and six touring cars, under an automobile escort of the various Coolidge-Dawes organizations, drew up in front of the City Hall at noon, where Mayor James Rolph Jr. bade the New England crusaders welcome.

Women Present

In the welcoming party of the Mayor were Mrs. Parker S. Maddux, representing the women's division of the Coolidge-Dawes organization, and Mark S. Requa, prominent Republican leader. Grouped about San Francisco's chief executive were hundreds of men and women, all active workers in the cause of Coolidge and Dawes.

Arriving with the transcontinental messengers of "Common Sense and Americanism" were A. W. Jeffries, former Congressman of Nebraska, and the man who placed the name of Charles G. Dawes in nomination for the Vice-Presidency, Captain C. E. Nusbaum of Sacramento, who is attached to the flying squadron of speakers of the State Central committee; Charles Doyle, secretary of the tour and a prominent business leader of Pittsburg; John P. Cowan, manager of the caravan and the head of the Tri-State Press of Pennsylvania, and a host of other prominent Republican workers.

LaFollette Hit

From an old platform on Grove street, which had been used as a reception stand for the World Flyers, mayor Rolph greeted the pilgrims, laughingly explaining that no formal welcome was possible within the City Hall, because it was the home of all the people, and was the one place in which politics had to be kept out.

In his speech Mayor Rolph decried the tendency of certain forces of labor to forsake the principles of the Republican party, and chase after the fantastic and uncertain theories of socialism under the banner of Robert M. LaFollette.

"The Republican party," said Mayor Rolph, "means for the working man prosperity, means at all times a good paying job, means a full dinner pail, means a happy and contented home, and why give up these things? No, no, I don't believe the San Francisco workingman is going to give up these things. On November 4, the workingman will be heard from, and on that day his cry as well as his vote will be for Calvin Coolidge and Charles G. Dawes."

Nebraskan Pleases

"Now if 'Bob,' I mean LaFollette wants a job, let him come to California. We'll give him work, but rest assured, we're not going to give him any such job President of the United States."

Former Congressman A. W. Jeffries, of Nebraska, a typical son of the West, caught the fancy of the crowd as soon as he moved to the front of the platform.

Picturesque with his six feet and four inches of brawn, in one hand a battered old black sombrero, and the other the state flag of Californaia, and hitting from right and left, the Nebraskan recounted just what the election of such a man as LaFollette would mean for the people of the United States.

"Take away your protective tariff," said Jeffries, "and what becomes of the American workingman? It means that he will have to compete with the cheap labor of a demoralized Europe – it means factories closing, it means banks failing, it means unemployment, it means money panics."

Woman Speaks

"The Republican party," continued the former Nebraska Congressman, "has accomplished many things during its last administration. It has collected quite a bit of the European war debt and is going after more. It has reduced the expenses of conducting a government by the budget system. All of these things mean lower taxes, and these are the things that make for contentment and prosperity."

Mrs. Parker S. Maddux, representing the women's division of the Coolidge-Dawes organizations, declared that the California woman was a combination of idealism and practical common sense. She told how her sex were anxious for life, liberty, home and pursuit of happiness, and of which revolved around the American citizen, man or woman, having a job.

"With the Republican party in power," said Mrs. Maddux, "every citizen has a job, and when this condition of affairs obtains then there is happiness and liberty – that's why eleven days from now the womanhood of California is going out almost to a woman to vote for Calvin Coolidge."

Sign Lincoln Register

Featuring the welcome to the Plymouth pilgrims was the signing of the Lincoln Register by those Californians who had the honor of casting a vote for the great emancipator. Those who signed the register were:

Charles P. Johnson, 86, formerly of Maine, 1423 Central avenue, Alameda; William Penn Willetts, 82, formerly of New York, 1926 Lake street, San Francisco; Morris Brinn, 86, formerly of Germany, 860 Sutter street, San Francisco; James Albert Dole, 81, formerly of Maine, Pomona, California; James L. Clark, 82, formerly of Iowa,266 Miramar avenue, San Francisco.

At the conclusion of the welcoming ceremonies the members of the Coolidge-Dawes Carivan were guests of the local Republican organization, at a luncheon at the Hotel St. Francis, following which they journeyed to Oakland, where they held a big rally.

Last night the caravan began its trip northward to Oregon and Washington."

A side article in the *San Francisco Examiner* featured a photo of the REO Speedwagons and an article on the vehicle:

Speedwagon in Long Caravan

An important part of the Coolidge-Dawes caravan, which checked into San Francisco yesterday after more than six weeks on the road from Plymouth, Vt., was played by a REO Speedwagon.

A fleet of passenger cars and the Speedwagon made up the caravan, which started from the town which claims the honor of being the birthplace of President Coolidge, for the purpose of visiting big cities and cross roads towns in a drive for votes for the Republican ticket.

The Speedwagon, with its sides carrying pictures of Coolidge and Dawes, served as a moving billboard as well as being a most effective carrier for printed literature and campaign pamphlets.

When the caravan reached this city it was welcomed by officials of the Reo Motor Car Company of California Inc., and a fleet of speedwagons all bearing the Reo emblem as well as Coolidge and Dawes signs preceded the visiting caravan down Market street.

Wilmer Wood, assistant manager of the Reo Motor Car Company of California, Inc., formally welcomed the visitors in behalf of the local company. The caravan before coming to San Francisco had visited the valley cities of Northern California then toured to Los Angeles and Southern California returning here via the coast route.

California Coolidge bumper sticker (14")

THE SAN FRANCISCO EXAMINER

Campaigners Forever

Here are the two oldest men to greet the Coolidge caravan on its arrival in San Francisco yesterday. They are (left) William P. Willits and Charles P. Johnson. Willits cast his first vote in 1864 and well remembers the "Tippecanoe" campaign. Johnson first went to the polls in 1857.

HISTORIC G.O.P. CARAVAN ARRIVES IN THIS CITY

Escorted by a brass band and followed by automobiles laden with officials of the local Republican organization, the Coolidge-Dawes Plymouth Caravan, consisting of motor trucks emblazoned with pictures of President Coolidge's birthplace in Plymouth, Vermont, and captained by ex-Congressman A. N. Jeffries of Nebraska, arrived in San Francisco yesterday morning.

At noon it was greeted in front of the City Hall by Mayor James Rolph.

The reception committee included Mrs. Parker Maddux, head of the women's Republican association; Mark L. Requa, chairman of the Republican Presidential State Committee; Charles L. Neumiller, chairman of the Republican State Central Committee, and Phil Katz, chairman of the Republican Association of War Veterans.

All members of the local committee spoke, but the principal address was delivered by Jeffries, who nominated Charles G. Dawes for Vice-President.

He said in part:

Our pilgrimage across thirteen States has been one triumphal procession. More than a million and a half people have heard what we have had to say about re-electing Coolidge. On the prairies, in populous cities and in the snowy regions of the Rockies we have found enthusiasm and a belief in the ultimate victory of the Republican party.

Mrs. A. B. Bradstreet, Mrs. Ed. Goeppert, Mrs. James Monks, Mrs. Edgar Glaser and Miss Daisy Lucey were among the leading Republican women in the crowd that assembled on the City Hall steps to welcome the standard bearers from Vermont.

The two oldest men in the assemblage were William P. Willits, who cast his first vote in 1864, and Charles P. Johnson, who voted for Lincoln in Alameda in 1857.

The two oldest men to greet the Coolidge Caravan on its arrival in San Francisco are William P. Willits (left) and Charles P. Johnson (right). Willits cast his first vote in 1864 and well remembers the "Tippecanoe" campaign. Johnson first went to the polls in 1857.

Saturday, October 25
Day 47

Left Sacramento, California at 2:00 p.m.

Log: No stops through Roseville, Marysville to Chico. 101 miles. Spent night at Oakes Hotel in Chico, California.
—E. A. Seidel.

Cornelius Vanderbilt Jr. Publisher

FINAL EDITION
TEL. HEMlock 3180

Illustrated Daily Herald

Weather Forecast
Today—Fair; W. winds. Temperatures yesterday: 5 a.m., 55.5; noon, 65.6; 5 p.m., 64.9. Tides: High, 9:27 a.m., 9:40 p.m.; low, 2:42 a.m., 3:24 p.m. Sun rises 6:27 a.m., sets 5:18 p.m. Moon rises 3:26 a.m., sets 4:12 p.m.

1¢

Vol. 1, No. 276—Twenty-eight pages — San Francisco, Saturday, October 25, 1924 — 1c a copy, street, and news stands; 5c on trains; 25c a month, $3.00 a year in California, Nevada. Elsewhere in U. S., 50c a month

—Illustrated Daily Herald Photo

SAN FRANCISCO GREETS COOLIDGE-DAWES CARAVAN—After traversing seventeen states and 6000 miles, the five automobiles, with fifteen occupants, which left Plymouth, Vt., the birthplace of President Coolidge, on September 9, arrived in San Francisco yesterday. At the left Mayor James Rolph Jr. is shown in the midst of the party greeting Charles P. Johnson, who came to California from Maine in 1857 and voted for Abraham Lincoln. The caravan truck is shown in the center, with President Coolidge's birthplace painted on the side. Standing before it is E. A. Seidel, who piloted it from New York. The photo at the right shows the caravan members exchanging greetings with reception committee members at the county line.

San Francisco Greets Coolidge-Dawes Caravan—After traversing 17 states and 6,000 miles, the five automobiles, with 15 occupants, which left Plymouth, Vermont, the birthplace of President Coolidge, on September 9, arrived in San Francisco yesterday. At the left Mayor James Rolph Jr. is shown in the midst of the party greeting Charles P. Johnson, who came to California from Maine in 1857 and voted for Abraham Lincoln. The Caravan truck is shown in the center, with President Coolidge's birthplace painted on the side. Standing before it is E. A. Seidel, who piloted it from New York. The photo at the right shows the Caravan members exchanging greetings with reception committee members at the county line.

Caravan parks in front of Civic Centre, San Francisco, California

Sunday, October 26
Day 48

Lay over at Chico, California

Log: Stayed here over Sunday; very quiet, good place to sleep. Rained all day.
—E. A. Seidel.

Hotel Oaks, Chico, California

Monday, October 27
Day 49

Left Chico, California at 8:30 a.m.

Log: Stops at Red Bluff, Redding, Dunsmier, Weed and Yreka. 187 miles. Spent night at Franco-American Hotel in Yreka, California.

Review: This was the second day of continuous rain. Dirt roads and very muddy.

Hotel here is the worst of the entire trip.

—E. A. Seidel

Tuesday, October 28
Day 50

Left Yreka, California at 9:00 a.m.

Log: Stops at Ashland, Medford, Central City, Gold Hill and Grant's Pass, all in Oregon. 92 miles. Spent night at Hotel Oxford in Grant's Pass, Oregon.

Review: Still raining. Roads all washed out as far as the Oregon line, where we once again hit the pavement.

Landslide caused delay of one hour, blocking the entire road.

Crossed the Siskiyou Peak, elevation 4,435 feet.

Flag presentation at Ashland, by local Republican leader.

—E. A. Seidel

Newspaper reports from Ashland and Medford follow:

Lincoln Party Welcomed and Receives State Flag

Ashland, OR – The Coolidge-Dawes Lincoln caravan was welcomed here today at noon and as a greeting received a beautiful Oregon state flag from Mrs. Louise Perozzi, republican state chairwoman and a local republican leader.

The caravan was two hours late in arriving here, the first stop in Oregon, having been delayed by a steam shovel working on the California part of the Pacific Highway and by a downpour of rain which retarded their progress over the mountains.

Despite inclement weather a large crowd gathered to hear A. W. Jeffris, a member of the caravan, who nominated Dawes at the Cleveland convention, extol Coolidge and Dawes and give many reasons why they should be elected.

Eleven pioneers who voted for Abraham Lincoln in 1860 and 1864 autographed the book which later will be presented to President Coolidge.

A. C. Spencer, local pioneer, formerly of Ohio, exhibited a framed union ticket ballot which was used in the last election of Lincoln and which he prizes highly.

Today marked the end of the eighth week of the caravan's trip across the continent.

Caravan at Medford

Medford, OR – The heavy rainfall of last night and today interferred with the elaborate plans to welcome to Oregon the Coolidge-Dawes automobile transcontinental caravan, yet the visit was a success in the way of stirring up republican enthusiasm. A. W. Jeffries, Nebraska, who nominated Dawes for vice-president at the Cleveland republican convention spoke at the rally.

AMERICAN POLITICAL ITEMS COLLECTORS

Collecting political campaign memorabilia, and studying political Americana has been called the "all-American hobby." Thousands of Americans collect campaign buttons, ribbons, banners, glassware & china, ballots and leaflets, books, autographs and photos. The United States is the world's oldest continuing democracy. This is an excellent way to learn about the issues and relive the significant events of America's political heritage...or collect the memorabilia associated with the social or political campaigns of today.

The American Political Items Collectors (APIC) is dedicated to the collection, preservation and study of materials relating to political campaigns. APIC was organized in 1945 and today has 2,000 members, many who are scholars of political history as much as they are collectors of the artifacts supporting that history.

Collecting and preserving the political memorabilia from these campaigns is important, but then researching and documenting the relevance completes the task.

Undoubtedly, if it had not been for the author's involvement in the APIC, the story presented in this book would never have been researched and written.

For more information about the American Political Items Collectors, or to become a member, write to: APIC Member Services, P.O. Box 55, Avon, NY 14414 or visit the Association's website at: apic.us

Wednesday, October 29
Day 51

Left Grant's Pass, Oregon at 8:30 a.m.

Log: Stops at Roseburg, Oakland, Yoncalla, Cottage Grove, Cresswell and Eugene, Oregon. 155 miles. Spent night at hotel in Eugene.

Review: More rain.

I. T. (Inner Tube Ole) Olsen of Portland now driving Lincoln.

Had dinner at the Rainbow Grill.

—E. A. Seidel

Roseburg coverage was reported in *The Morning Oregonian.*

Caravan Visits Roseburg. Public Demonstration Prohibited by Downpour of Rain

The Coolidge-Dawes Lincoln caravan arrived here at noon today. Because of the hard storm the caravan was delayed in reaching the city from Grant's Pass, being more than an hour behind schedule. Owing to the weather no public demonstration was possible, and the party stopped only for a short meeting and luncheon.

Representative Jefferis outlined the political outlook as investigated by members of the caravan during their tour to the Pacific coast, predicting the election of Coolidge. He gave facts and figures showing the great increases in taxation, which would result should government ownership of railroads, as proposed by LaFollette, be adopted.

Following the meeting at the city hall the members of the party were entertained at a luncheon at the Umpqua hotel, where a few short talks were made by local committee members and those in the caravan.

As no demonstration was possible, because of the downpour, the caravan left early in the afternoon, stopping at Sutherlin, Oakland, Yoncalla and Drain on the way north to Eugene, where the stop was made for the night.

Thursday, October 30
Day 52

Left Eugene, Oregon at 9:30 a.m.

Log: Non-stop to Portland. 128 miles. Spent night at Hotel Imperial in Portland.

Review: The old Studebaker breathed her last at eugene this morning and was left behind.

Rained all day; cancelling all meetings.

—E. A. Seidel

The *Morning Oregonian* on Thursday carried this feature on the caravan's upcoming visit to Portland:

Caravan To Reach Portland Tonight. Parade Will Meet Tourists at Oregon City. Mass Meeting is Billed. M. E. Crumpacker, Republican Candidate for Congress, to be Chairman at Gathering.

The Coolidge-Dawes Lincoln caravan, consisting of eight automobiles and one truck, the latter bearing republican literature, will arrive at Oregon City this afternoon at 5:30 o'clock. The caravan will be met by about 100 automobile loads of Multnomah county republicans, according to John L. Day, county chairman, who is arranging the Multnomah county reception at Oregon City and the mass meeting to be held tonight at 8 o'clock in the Pythian temple auditorium, West Park and Yamhill streets.

M. E. Crumpacker, republican nominee for congress from Multnomah county, will be marhall of the parade and he will also be chairman of the mass meeting, introducing the speakers. W. J. Jefferis, ex-representative in congress from Nebraska; Herbert L. Moore, one of Coolidge's boyhood farmer friends; Colonel Thomas R. Hamer, ex-representative in the congress from Idaho, republican state committeeman from Multnomah county, and Miss Elinore Barker will be the speakers.

Miss Barker to Speak

Miss Barker is a prominent attorney of Indianapolis, Ind., who is making a hurried tour of the west under auspices of the national committee. After speaking at several women's luncheons she consented to remain over in Portland for this mass meeting. Alice Price Moore, contralto, will lead in community singing of "America," accompanied by Mrs. Helen Van Houten, pianist.

The caravan was originated during Mr. Coolidge's vacation at the farm home of his father in Vermont and left Plymouth September 9. Oregon will be the 16th state covered by the boosters. The cars carry banners and flags presented to them by Coolidge organizations in the different states, and probably the caravan's proudest possessions are the state flags, including a flag carrying the insignia of the Beaver state, presented as the group reaches the border of each state, and the registration book in which pioneers who voted for Lincoln are entitled to register. The book will be suitably bound and presented to President Coolidge when the caravan disbands.

Parade Route Given

It is estimated that at least 60 old-time republicans who voted for Lincoln will join the parade to Oregon City. Those who have signed up to go are asked to meet at the Fourth street entrance to the courthouse this afternoon at 4:45 o'clock, where cars, the use of which have been donated by local automobile dealers, will await them.

The parade will form at the south park blocks, the head of the line to be at Jefferson street, to proceed at 4 o'clock to Oregon City on the westside highway. The line of march on the return to Portland will be: After crossing the Clackamas river at Oregon City, the East Eighty-second street route to Foster road, west on Foster road to Fifty-second street, north on Fifty-second street to Hawthorne avenue, west on Hawthorne to Grand avenue, north on Grand to Oregon street, crossing the steel bridge to Third street, south on Third to Morrison, west on Morrison to Tenth, north on Tenth to Washington, east on Washington to Broadway and north on Broadway to the Imperial hotel entrance to disband. The headquarters of the visitors will be the Imperial.

Large 4" Coolidge "Deeds - Not Words" celluloid easel button

Another Portland newspaper carried the following story:

Auto Caravan Due in City Thursday. Coolidge-Dawes Party Covers 6000 Miles. 15 States Traversed. Men Who Voted for Abraham Lincoln Register Names and and Present Book to Travelers

The Coolidge-Dawes automobile caravan, which traveled nearly 6000 miles across the continent from Plymouth, Vt., will stop in this city on the auxillary tour embracing the Pacific coast states. The caravan is scheduled to arrive here for a meeting Thursday night.

Leaving the birthplace of President Coolidge on September 9, the little caravan of six cars has been kept intact, the same drivers making the entire journey to the coast. This has been the longest procession in history. Fifteen states have been traversed and more than 350 meetings have been held with an aggregate attendance of 300,000 voters. It is estimated that 5,000,000 persons have witnessed the progress of this campaign parade, taking part in the receptions or in cheering the travelers on their way.

100,000 Autos Participate

More than 100,000 automobiles have participated in the motor escort furnished from county to county in the long trip across the country. At times these processions were 15 miles long. The route covered southern Vermont and New Hampshire, western Massachusetts and central New York, and after leaving New York City, followed the Lincoln highway to San Francisco. More than a dozen mountain ranges were crossed, including the Berkshires and Catskills, the Alleghanies, the rockies and the Sierras. In making their way through Wyoming a foot of snow on the Laramie range failed to baffle the intrepid drivers. In one mountain community an escort of cowboys on ponies led the way for the mud-covered automobiles.

The caravan carries the flags of 15 states presented by the women's committees of each state crossed on the way. The flags will be presented as mementoes of the journey by the republican national committee. They will be displayed at the meeting here.

Distinguished Men in Party

The officers of the caravan are: A. W. Jefferis of Omaha, Neb., director and principal speaker; John P. Cowan, Pittsburg, Pa., manager, and Charles T. Doyle, Pittsburg, secretary. The drivers of the cars, all graduate automotive engineers, are: E. A. Seidel, Ernest Hoch and S. E. Mollden, New York, and William Wallace and A. W. Jefferis Jr, Omaha. Mr. Jefferis at the Cleveland convention placed the name of General Charles G. Dawes in nomination for the vice-presidency and made the official notification speech at Evanston. He formerly served in congress from Nebraska. Several other distinguished speakers will accompany the caravan through Oregon and Washington.

When the party left Plymouth a group of aged men who had voted for Abraham Lincoln and who knew Calvin Coolidge as a boy wrote their names in a book and presented it to the travelers, asking that all men found in route who likewise had voted for the first republican president be enrolled. More than 1000 Lincoln voters have been enrolled and the book will be rebound and presented to President Coolidge. Among those who registered are several who voted for John C. Fremont in 1856 and who have the distinction of having voted for every presidential candidate of the republican party.

Veterans Invited to Register

Men who voted for Abraham Lincoln who wish to participate in the parade to Oregon City tomorrow afternoon may register either at republican headquarters, 222 Oregon building or at G.A.R. headquarters in the courthouse today before 5 p.m. The veteran republicans will meet at the Fourth street side of the courthouse at 3:45, where automobiles, the use of which have been donated by local dealers, will pick them up and join the parade at the south park block, from where the start for Oregon City will be made. The head of the parade will be formed at Jefferson street.

The caravan will arrive in Oregon City on its way north about 5 p.m. A mass meeting is to be held at Pythian temple, Portland, West Park and Yamhill streets, at 8 p.m. W. A. Jefferis of Nebraska, Herbert L. Moore of Vermont and Colonel Thomas R. Hamer of Portland will be speakers. Alice Price Moore, contralto, will sing accompanied by Mrs. Helen Van Houten. M. E. Crumpacker will be marshall of the parade and he will also serve as chairman of the mass meeting.

A Portland newspaper article reporting after the event read as follows:

Coolidge Caravan Reaches Portland. Trip One Long Ovation, Members Declare. Portland Republicans Turn Out in Force and Give Visitors Enthusiastic Welcome

Near the end of what they say has been practically a continuous procession and ovation all the way from President Coolidge's hometown of Plymouth, Vt., members of the Coolidge-Dawes Lincoln tour automobile caravan reached Portland yesterday.

Members of the party, who have spoken in behalf of Coolidge and Dawes at some 400 meetings since the caravan left Plymouth September 9, are A. W. Jefferis, John P. Cowan, Pittsburg, manager of the caravan, and Charles T. Doyle, Pittsburg, secretary of the caravan.

In the party also are the following campaign workers who have piloted the caravan across the country: E. A. Seidel, Ernest Hoch and S. E. Mallden, New York City, and William Wallace and A. W. Jefferis Jr of Omaha.

"If the time every comes when, as is advocated by Robert M. LaFollette in his candidacy for president of the United States, the supreme court of the United States is so placed that it can't keep congress within bounds of the American constitution, then the liberties of the American people will soon pass away."

Such was the declaration pounded home last night by A. W. Jefferis of Omaha, Neb., ex-representative in congress and member of the caravan, at an enthusiastic meeting at the Pythian temple, West Park and Yamhill streets.

It was Mr. Jefferis who placed Charles G. Dawes in nomination for the vice-presidency on the republican ticket at the national convention in Cleveland, and in urging the election of the republican candidates next Tuesday Mr. Jefferis paid high tribute to Mr. Dawes, reviewing his war work, the successful culmination of his reparations plan for the rehabilitation of Europe, and his status as a man of high moral courage and character, his patriotism, running back through generations of patriots, and his broad conception of his duty to his country and to his fellow man.

Concluding his plea for the election of the republican ticket next Tuesday Mr. Jefferis said:

"If the American people do not want to face a crisis of national chaos the only thing for them to do at the polls next Tuesday is to elect Calvin Coolidge and Charles G. Dawes as president and vice-president of the United States.

"If they do not elect Mr. Coolidge they will elect no one. Mr. Davis will carry a few states. Mr. LaFollette will carry a few states. Neither one of them can be elected. Their combined strength, however, if enough votes are given them, might prevent the election of Mr. Coolidge.

"In that event the American people have tied their own hands. Chaos will follow. No one can say what the results may be."

Mr. Jefferis reviewed the peace and prosperity of the United States as it has been spread before members of his party on their transcontinental automobile caravan. He reviewed the early struggles and...

Friday, October 31
Day 53

Left Portland, Oregon at 9:00 a.m.

Log: Stops at Vancouver, Woodland, Kalama, Kelso, Longview, Castle Rock, Toledo, Centralia and Chehalis, all in Washington. 113 miles. Spent night at Hotel St. Helene in Chehalis, Washington.

Review: Congressman Jefferis, Jeff, Jr, and Bill Wallace left us this morning at Portland. Only five of us left that started from Plymouth: Cowan, Doyle, Hoch, Mollden and myself. It sure is quiet without the bunch.

Hoch has a bad attack of ptomaine poisoning and is being left in the hotel with Doyle.

Mollden had two flats inside of ten miles.

Dinner at Centralia Hotel by Republican Club.

—E. A. Seidel

Hotel St. Helens, Chehalis, Washington

A Centralia newspaper report follows:

Caravan To Be Given Welcome by Republicans. Meeting to be Held at High School Auditorium Tonight in Honor of Party that is Finishing Cross-Country Tour for Coolidge and Dawes.

Members of an automobile caravan that is completing a cross-country tour from Plymouth, Vt. to Bellingham, Wash., in the interests of the Coolidge-Dawes candidacy will be honor guests at a mass meeting to be held at the high school auditorium this evening by local republicans.

The caravan is due to arrive in Centralia about 5 p.m. A big delegation of cars left here early this afternoon for Castle Rock to meet the tourists and act as an escort of honor to this city. Congressman Albert Johnson was included in the escort party and is scheduled to be one of the speakers this evening.

W. H. Cameron, local attorney, will preside at tonight's meeting, which all republicans are urged to attend.

Near the end of what they say has been practically a continuous procession and ovation all the way from President Coolidge's "home town" of Plymouth, Vt., members of the Coolidge-Dawes caravan reached Portland yesterday.

Members of the party, who have spoken in behalf of Coolidge and Dawes at some 400 meetings since the caravan left Plymouth September 9, are A. W. Jefferis of Omaha, ex-representative in congress, who placed the name of Charles G. Dawes in nomination for the vice-presidency at the Cleveland convention; John P. Cowan, Pittsburg, manager of the caravan, and Charles T. Doyle, Pittsburg, secretary of the caravan.

In the party also are the following campaign workers who have piloted the caravan across the country: E. A. Seidel, Ernest Hoch and S. E. Mallden, New York City, and William Wallace and A. W. Jefferis Jr., of Omaha.

Herbert L. Moore, one of President Coolidge's boyhood friends, left the caravan at Sacramento, that he might be home in time to vote and Senator Woodsworth of New York, who also started with the caravan, went back to manage the gubernatorial candidacy of Theodore Roosevelt, Jr. in New York.

Mr. Cowan said in Portland that the caravan had traveled practically 6,000 miles and that it had been received enthusiastically all along the line. At least 100,000 automobiles, he said had participated in the various processions held en route, and in Indiana an airplane followed the caravan overhead distributing campaign literature across the state.

Mr. Jefferis left for the East last night after his Portland address. Others in the party will take their time about getting home from Bellingham. They have arranged, Mr. Cowan said, to cast their votes by mail.

The Coolidge-Dawes Lincoln tour was conceived by the boyhood friends and neighbors of President Coolidge all residents of Plymouth and its environs.

A welcoming editorial by Harry L. Bras is simply entitled **"The Coolidge-Dawes Caravan"**

Centralia feels complimented by your presence here today and we trust our citizens will fully appreciate the sacrifices you have made in undertaking this unique pilgrimage across the continent in behalf of Coolidge and Dawes candidates for the highest office in the gift of the American people.

You have taken the trail, as it were, to do your part in saving this nation from the terrible consequences of radicalism. You have chosen to travel on the transcontinental highway across the continent named for the Great Emancipator. You have wisely chosen the Lincoln Highway in memory of the great man who freed the slaves at a time when slavery was the paramount issue, and we believe now that we have a right to invoke the spirit of Lincoln in behalf of the greatest issue in the campaign, that of freeing this nation of the menace of radicalism – a menace far more dangerous in the fundamental principles of this nation than was slavery in 1860.

Starting as you did at the humble birthplace of Calvin Coolidge you are bearing this personal message to all who are fortunate to "live by the side of the road" and receive the inspiration of your visit. WE WELCOME YOU.

Official Caravan line-up in Portland, Oregon

Saturday, November 1
Day 54

Left Chehalis, Washington at 9:00 a.m.

Log: Stops at Tenino, Olympia, South Tacoma, Puyallup and Tacoma. 100 miles. Spent night at Hotel Tacoma in Tacoma, Washington.

Review: Rain again all day.
Staying here over Sunday. Olsen and I going to Seattle.
Governor Hart and staff, also Congressman Johnston accompanying us.
Hock and Doyle rejoined us here Saturday night.
Ted Brandberg of the Portland Dodge Co. now driving Wallace's car.
—E. A. Seidel

Sunday, November 2
Day 55

Lay over at Tacoma, Washington.

Monday, November 3
Day 56

Left Tacoma, Washington at 8:00 a.m.

Log: Stops at Auburn, Sumner, Seattle, Everett, Mt. Vernon and Bellingham. 165 miles. END OF TRIP.

Review: And still it rains.

Big parade and meeting here at the Rialto theatre.

Governor-elect Hartley was one of the speakers here.

J. P. Cowan in final speech gives the drivers and their cars credit for the success of the trip, and introduces us from the stage.

Sold the truck, minus the body to the Reo dealer at Everett.

Well, its all over now, but it was a great trip.

What is left of the original caravan. J. P. Cowan, Charlie Doyle, Ernest Hoch, Elof Mollden and E. A. Seidel. The Reo and the two Dodges are the cars that completed the trip.

—E. A. Seidel

The daily mile average was 125 miles, the shortest jump, 65 miles, the longest, 195 miles.

The caravan passed through Vermont, New Hampshire, Massachusetts, New York, New Jersey, Pennsylvania, Ohio, Indiana, Illinois, Iowa, Nebraska, Wyoming, Utah, Nevada, California, Oregon and Washington. By the time the caravan reached its final destination it was carrying the flags of 17 states.

What is being extolled as one of the most remarkable documents ever compiled is the Lincoln Register, compiled along the Coolidge-Dawes Lincoln Tour. At each stop the register was signed by men who had voted for Abraham Lincoln 64 years before.

The Bellingham paper carried the following note:

CHIEF THANKS CARAVAN.

Coolidge-Dawes Tour Ended at Bellingham, Wash.

John P. Cowan of Pittsburg, Pa., manager of the Coolidge-Dawes Lincoln caravan of six automobiles, bearing friends and neighbors of President Coolidge, of Plymouth, Vt., today received a telegram from the president in which he expressed his appreciation of the manner in which the tour was conducted.

The caravan ended its trip of 6500 miles through 17 states with a demonstration here last night.

The finish line. A little dirty, but still going strong.

Tuesday, November 4
DAY 57

ELECTION DAY

Calvin Coolidge and Charles Dawes won by a landslide with a popular vote more than their two primary rivals, John W. Davis and Robert M. LaFollette, combined. In the popular votes, Coolidge received 15,725,016 compared to the Democrat Davis with 8,386,503 and the Progressive LaFollette with 4,822,856. Coolidge received 382 electoral votes, Davis 136 and LaFollette 13.

NAMES AND ADDRESSES of All Subscribers OPEN TO INSPECTION

BATTLE CREEK MOON-JOURNAL

THE HOME NEWSPAPER In a Community of OVER 40,000 PEOPLE

VOLUME 10, NUMBER 128 — BATTLE CREEK, MICHIGAN, WEDNESDAY AFTERNOON, NOVEMBER 5, 1924. — PRICE THREE CENTS

COOLIDGE IS EASY WINNER

PRESIDENT SWEEPS NEARLY EVERYTHING BUT "SOLID SOUTH"

Is Given Estimated Electoral Vote of 379; Bob La Follette Wins Only in His Home State.

By ROBERT J. BENDER
United Press Staff Correspondent

New York, Nov. 5.—Coolidge, 379; Davis, 139; La Follette, 13.

With the possible exception of a switch of New Mexico's three votes from Davis to Coolidge, this will be the standing of the Presidential candidates in the electoral college when final tabulations are in.

In New Mexico, where the count always is slow, the Democrats waged a strong fight against "Fall rule" and the stigma imposed upon the former secretary of the interior by the oil scandal was expected on the face of early returns to defeat President Coolidge in the state.

Of the other doubtful commonwealths, Missouri turned definitely into the Coolidge column today after wavering throughout the night; and in Montana, where Senator La Follette was making a strong bid for the electoral votes, the day's tabulations showed steadily

Shows Electoral Vote on Tuesday

State.	R.	D.	I.
Alabama		12	..
Arizona	3	1	..
Arkansas		9	..
California	13	..	..
Colorado	6	..	..

WINS SWEEPING VICTORY

FIRST WOMEN ARE ELECTED AS GOVERNORS

Mrs. Ferguson Wins Texas Chair and Mrs. Ross in Wyoming.

AL SMITH IS RE-ELECTED

Overcomes a Big Coolidge Lead; Vic Donahey is Ohio Winner Over Harry Davis.

New York, Nov. 5.—(U. P.)—Two women governors, the first in the history of the republic, have been elected—Mrs. Miriam A. ("Ma") Ferguson in Texas and Mrs. Nellie T. Ross in Wyoming.

Mrs. Ferguson, outspoken enemy of the Ku Klux Klan, in a Klan stronghold, and wife of James E. Ferguson, who was impeached from the governorship for alleged misuse of the state funds, staged one of the most dramatic campaigns of the election.

After she was named the Democratic nominee, the Republicans sought to prevent her election by obtaining an injunction against her on the grounds that it was against the state constitution for a woman to be elected to the governorship. Mrs. Ferguson won the court fight and then the election by a comfortable majority over her Republican opponent.

Mrs. Ross, also a Democrat, is

They Won Seats in U. S. Senate

New York, Nov. 5.—(U. P.)—The following Senators were elected:

Alabama—Thomas J. Heflin, Democrat.
Arkansas—Joseph T. Robinson, Democrat.
Colorado—(Incomplete).
Delaware—T. Coleman Dupont, Republican.
Georgia—William J. Harris, Democrat.
Idaho—William E. Borah, Republican.
Illinois—Charles E. Deneen, Republican.
Iowa—(Incomplete).
Kansas—Arthur Capper, Republican.
Kentucky—Fred M. Sackett, Republican.
Louisiana—Joseph E. Ransdell, Democrat.
Maine—Bert M. Ferndale, Republican (Elected September).
Massachusetts—(Incomplete).
Michigan—James E. Couzens, Republican.
Minnesota—(Incomplete).
Mississippi—Pat Harrison, Democrat.
Montana—(Incomplete).
Nebraska—George W. Norris, Republican.
New Hampshire—Henry W. Keyes, Republican.
New Jersey—Walter E. Edge, Republican.
New Mexico—(Incomplete).
North Carolina—Furnifold M. Simmons, Democrat.
Oklahoma—W. B. Pine, Republican.
Oregon—Charles L. McNary, Republican.
Rhode Island—Jesse H. Metcalf, Republican.
South Carolina—Coleman L. Blease, Democrat.
South Dakota—W. H. MacMaster, Republican.
Tennessee—L. D. Tyson, Democrat.
Texas—Morris Sheppard,

COUNTY GIVES BIG VOTE TO REPUBLICANS

La Follette Runs Second in Battle Creek in the Presidential Race.

WILLIAMS WINS EASILY

Henry, Curtis, Couzens and All Other G. O. P. Candidates Are Winners in Calhoun.

Calhoun county, casting the heaviest vote in its history, went overwhelmingly for Coolidge and all other Republican candidates at yesterday's general election. The President was given a five to one vote over John W. Davis, the Democratic candidate for President who ran second, and a six to one vote over La Follette, who ran third.

Practically complete county returns this afternoon showed the vote for President as follows:

Coolidge, 16,834.
Davis, 3,864.
La Follette, 2,777.

Seventeen of Battle Creek's eighteen precincts contributed 8,471 of Coolidge's county vote, while La Follette ran second in the city with 1,912 votes and Davis third with 1,527 votes. A heavy railroad vote here was responsible for La Follette's showing, election officials said.

Acknowledgements

The author would like to extend appreciation to the following individuals, groups and institutions for their assistance with the development of this book.

Allen County Museum, Allen County Public Library, American Political Items Collectors, Aurora Public Library, Calvin Coolidge Memorial Foundation, Catskill Mountain News, Chicago Heights Public Library, Conococheague District Library, Cortland Free Library, Everett Free Library, Goodall City Library, Goshen Public Library, Greensburg Area Library, Holyoke Public Library, Huntington Memorial Library, Ilion Free Public Library, Keene Memorial Library, Kern County Library, Kingston Area Library, Lincoln Highway Association, Lincoln Highway National Museum, Lodi Public Library, Los Angeles Public Library, Madera County Library, Marshalltown Public Library, Mary Lou Johnson/Hardin County District Library, Newark Public Library, Poughkeepsie Greater Library District, Rensselaer County Historical Society, Richards Free Library, Rick Moses/Americana Resources, Rock Springs Historical Museum, Roxbury Library Association, Salem Public Library, Santa Clara County Historical Society, St. Joseph County Public Library, Sterling-Rock Falls Historical Society, Tioga County Historical Society, Tribune-Review Publishing Co., Utica Public Library, Warner Library, Washoe County Library, Yonkers Public Library